The Space Shuttle

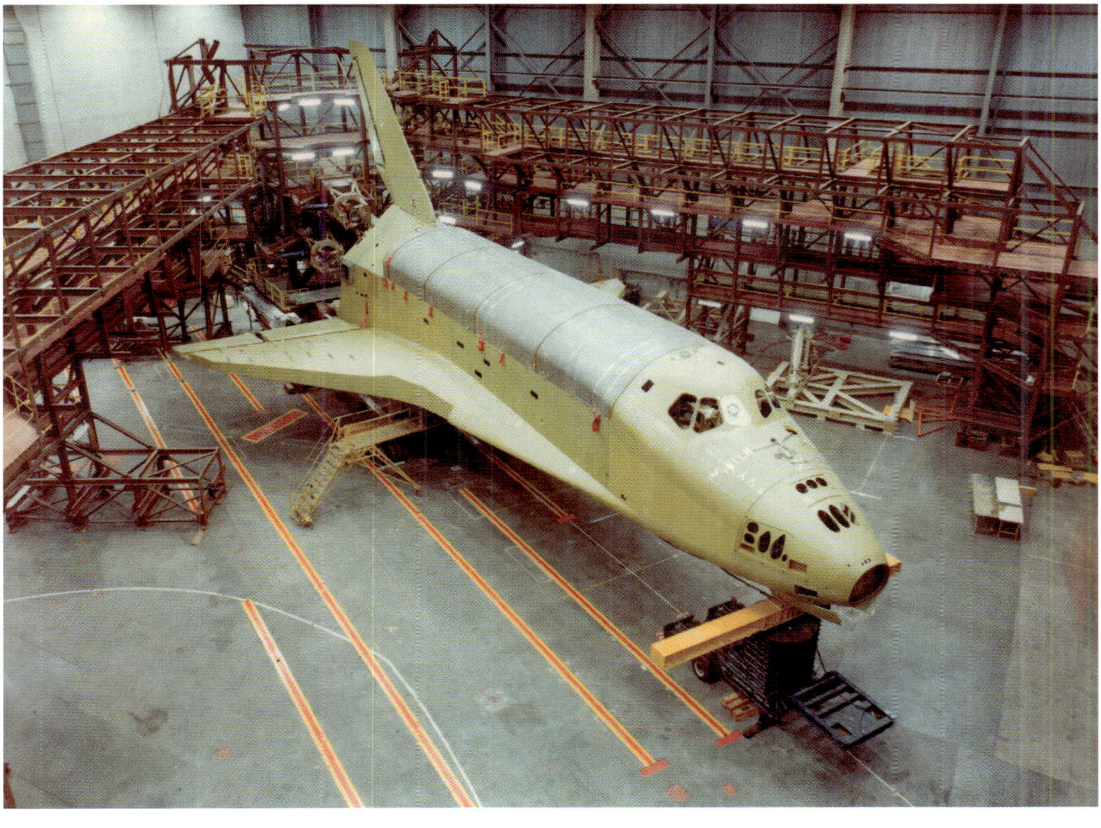

DAVID BAKER

KEY Books

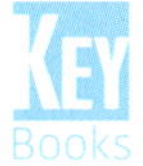

Published by Key Books
An imprint of Key Publishing Ltd
PO Box 100
Stamford
Lincs PE9 1XQ

www.keypublishing.com

Typeset by SJmagic DESIGN SERVICES, India.

Contents

Introduction

During the 135 launches between 1981 and 2011, the National Aeronautics and Space Administration's (NASA) reusable transportation system – the Space Shuttle – launched and serviced many satellites, space-based observatories, and planetary missions. Furthermore, it lifted into orbit most of the elements for the giant, 400-tonne International Space Station (ISS), which has supported a permanent human presence in space since the year 2000.

The Shuttle has been criticized for being too expensive to develop, not fulfilling the expectations of engineers who conceived it and the politicians who funded it, being expensive to fly, and flawed for having little margin for crew safety. On that latter point, two missions failed to return home with their crew; a total of 14 astronauts losing their lives in 1986 and 2003.

Yet, despite all of this, it was undeniably a magnificent engineering achievement, conceived in a time of great optimism and change, both in the space program and the world at large. It helped build a commercial space industry by launching non-government satellites for private companies and helped project American aerospace capabilities to a global market, bringing peripheral benefits from cooperative agreements and all manner of ventures, linking countries around the world. Above all else, it pushed European governments and organizations to come together, culminating in the establishment of the European Space Agency (ESA). This, together with cooperation from Canada and Japan, forged the partnership that made the ISS possible and brought in a post-Soviet era Russia as a fourth partner. Without the lift capacity of the Shuttle, the ISS would not have been possible, as eventually more elements were assembled from manufacturers outside of America than inside.

These achievements are perhaps the greatest legacy of the Space Shuttle system. Although its last flight was more than ten years ago, what it made possible lives on ever more productively. Conceived 65 years ago, before the first Apollo spacecraft had been launched, and two years before American astronauts walked on the Moon, as an engineering marvel the Space Shuttle was the greatest winged flying machine of all time.

Orbiter *Atlantis* introduces a state-of-the-art glass cockpit for STS-101. (NASA)

Chapter 1

A New Beginning

I t was not altogether coincidental that, in 1967, the year NASA lost three astronauts in a tragic fire with the first manned Apollo spacecraft, the agency began a transformation that would focus on winged space vehicles plying back and forth between the ground and low Earth orbit. Gone would be the ballistic capsules with the appearance of hollowed out warheads, carrying astronauts on brief rides into space and back to a splashdown in the Atlantic or Pacific oceans.

The fire that took the lives of Grissom, White and Chaffee in a pad rehearsal for the first manned Apollo flight in January 1967 was a grim warning of dangers that the space agency could never fully avoid. Safety precautions were fundamental to the design of the one-man Mercury, the two-man Gemini, and the three-man Apollo, but haste to get to the Moon by the end of that decade cut corners, dodged critical quality-control precautions, and accepted risk as an inherent price for progress.

Bound by the focused decision to get astronauts on the Moon before the Russians – never as close-run as many thought at the time – NASA had expanded far beyond expectations when it was established in October 1958 in response to the Soviet sputniks. Single-minded and with costly new

The Shuttle program completed 135 flights between 1981 and 2011, but the configuration that became familiar to people around the world was originally conceived as a very different vehicle. (NASA)

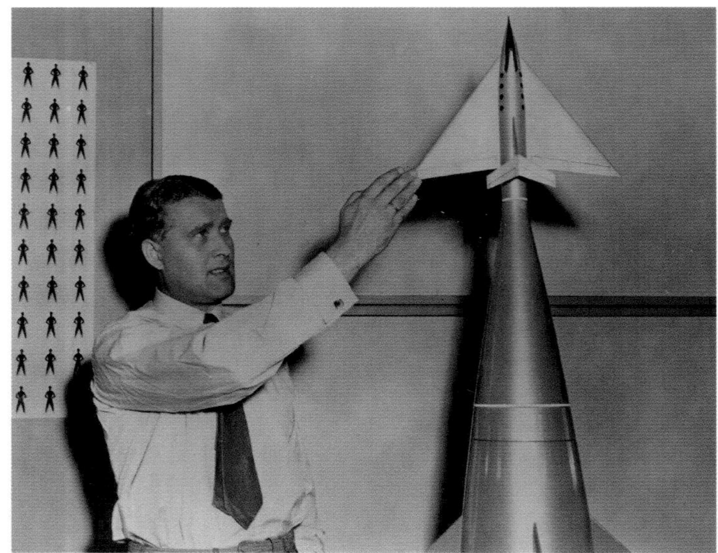

The idea to build a reusable space vehicle originated in the 1950s, when most people, including the German rocket scientist Dr Wernher von Braun, who directed the V2 missile during World War Two and the development of the Saturn rockets for NASA's Apollo program, believed that space flight would be achieved through winged craft launched by conventional rockets. (US Army)

rockets and space vehicles as core elements of that capability, it became apparent when the budget began to slide into decline that reusability would have to trump expendability if the program was not only to survive but also grow.

While greatly improved Apollo capsules were crafted and extensively checked, tests were completed with the giant Saturn V rocket. Initial flights were made from 1966 with the much smaller Saturn IB for the first manned Apollo flight and a test run of the Lunar Module, a bespoke vehicle designed to carry two astronauts from lunar orbit to the surface. By 1968, during a year in which the first Apollo would rise like a phoenix from the ashes of a once-flawed vehicle, NASA made the commitment to develop a vehicle capable of carrying up to seven astronauts and a hefty payload to low Earth orbit (LEO).

Success with the Saturn rockets hinged on the reliable performance of cryogenic engines, rocket motors burning liquid oxygen (LOX) and liquid hydrogen (LH2). The significant improvement in power/weight ratio provided by those propellants meant that it would be possible to achieve the relatively heavy lift needed to do the required work, and this also made a Space Shuttle possible. But sheer power and lift capacity were only two of several technology hurdles to be overcome.

Even with cryogenic propulsion and all the other necessary technologies in place, it was not possible to get into orbit with a single rocket stage. The laws of physics and the limitations of material technology could not provide a sufficiently light structure to allow that. At least two rocket stages were necessary. Some hope came with the development of the North American X-15, which completed 199 free flights between 1959 and 1968. Ironically, the X-15 had been developed as a stepping stone to space flight with winged vehicles. The US Air Force (USAF) had such a project, the Dyna-Soar (for "dynamic soaring"), an unfortunately prophetic name for a concept felled by a presidential axe in 1963 after NASA received the Moon challenge.

But the X-15 was not a dead end. The ability to propel a reusable vehicle into space and get it back intact relied on several things: energy management; a major research program within which the X-15 would provide data and pilot experience; and a starting point for electronic and solid-state electronic flight control systems that would underpin Shuttle technology. In that regard, lift through powered flight both into and out of space inspired additional programs; for instance, to minimize the amount of heat from re-entry, various aero-shapes were tested through NASA's Lifting Body program, in which three very different blended wing-body designs were flown and tested during a 12-year-long program beginning in 1963.

On the road to some form of shuttlecraft, by 1968 the decision had been made to combine a launch vehicle and space vehicle in a single entity. On October 30, NASA asked the aerospace industry to submit proposals for what it called an Integral Launch and Re-entry Vehicle (ILRV). The driving force was the target for low cost, and the request came not only days after the successful return to Earth of the first manned Apollo flight but also as the NASA budget was taking a nose-dive. Irrevocably committed to getting astronauts on the Moon within the next year, NASA judged that funds projected for the early 1970s would be insufficient to mount a post-Apollo program using these expensive, expendable rockets and spacecraft. Reusability and the economics promised by the shuttlecraft concept seemed to be the only solution.

Right: By the late 1960s, NASA was heavily committed to landing astronauts on the Moon using the giant Saturn V rocket, which was costly and could be used only once. (NASA)

Below: NASA astronauts (from left) Gus Grissom, Edward White II, and rookie Roger Chafee lost their lives in a fire on January 27, 1967, while rehearsing the first manned Apollo flight, an event with consequences that changed the way the future would unfold. (NASA)

A redirection of effort

The decision to pursue an ILRV was a profound move by NASA, with the implication greatly at variance with its wider image, which was to explore and send spacecraft to deep-space destinations. Publicly, in 1968, NASA had broadcast a message in which it promised repeated Moon flights and manned missions to Mars as an achievable goal, certainly by the end of the century. But the reality was very different. The Shuttle would be incapable of delivering anything above LEO and certainly not in support of any further Moon missions. It would be a space delivery truck, with the intention of supporting the assembly of a space station.

In January 1969, Richard Nixon entered office as the 37th President of the United States and immediately challenged NASA's pre-eminent position in national priorities. Vice President to Dwight D. Eisenhower until January 1961, Nixon had lost none of the cynicism from his boss over manned flight and grand goals. That view persisted, despite him using the Apollo 11 mission to bolster his own presidency and basking in the global attention gravitating to Armstrong and Buzz Aldrin as they undertook a worldwide tour following the Moon landing in July 1969. There were fears among many that Nixon would shut down manned flight, seeing it as having won the space race. In fact, he did contemplate transforming NASA into a general national technology agency for all manner of science and engineering challenges.

Nevertheless, in February 1969, initial design proposals were requested on the ILRV and General Dynamics/Convair, Lockheed, McDonnell Douglas, and North American all submitted proposals. It was necessary to get as wide a range of support for the ILRV as possible. At this date, the USAF was launching a lot of satellites, and with the prospect of replacing expendable rockets with a reusable vehicle, it too requested studies from the same companies. But there were problems. NASA had decided on a payload figure of 11,340kg (25,000lb) for its own manifest, but with very heavy spy satellites in development, the USAF wanted twice that capability.

Despite the negative view from the White House, NASA had many friends in Congress, the agency having done a superb job distributing field centers and selected contractors across the nation, bringing jobs and local support. But it also needed to show that other potential customers for load-carrying missions existed. After all, the more flights there were, the less each one would cost. Seeing in the USAF a major supporter on which it could rely when asking Congress for money to build it, on May 5, 1969, NASA asked the ILRV contractors to look at a vehicle capable of carrying 22,680kg (50,000lb) of payload to LEO.

In September 1969, NASA presented a future in which a space station would be serviced by the Shuttle, and a Space Tug would be developed to move satellites between low and high Earth orbits,

Pulling away from the chase-plane observing its release from a B-52 carrier-plane, the North American X-15 accelerates to hypersonic speed under the power of its throttleable rocket motor. The X-15 was considered a precursor step to orbital space flight using winged vehicles. (NASA)

The US Air Force (USAF) developed the Dyna-Soar spacecraft as the first step on a path to winged aerospace vehicles capable of suborbital and orbital missions, including offensive operations. (USAF)

servicing them for longer life. Also proposed was a nuclear upper stage launched by two-stage Saturn V rockets to cut transit time between Earth and between the Moon and Mars, using the former as a springboard to the latter. The plan got the support of Vice President Spiro Agnew, but it quickly became clear that there would be insufficient money to produce the Shuttle and the station in parallel. Since the station needed a Shuttle to keep it supplied, that would be developed first, followed by the station when the country could afford it. There never was any money for the Space Tug or the nuclear upper stage

By the end of the year, the budget was still falling, and NASA was running out of money. With most of the development and hardware production completed, allocations for Apollo were coming down. But these savings were not being replaced with funds for a new and more economical way of running a space program. In fact, Nixon even attempted to cancel the last three Moon missions, already cut back by three flights as a budget-saving bid. For these reasons, NASA consolidated its future space plan with the Shuttle at its core. To do that, it needed a credible development plan and a convincing strategy for political approval.

It would achieve that through a phased program, taking the various optional alternative designs and moving them through critical analysis prior to filtering out the best. Phase A was a feasibility study, considering the concept and testing its viability with recourse to a specific design. Phase B would define specific design proposals and probably involve two competing corporate contenders. Phase C would be the detailed design from a company or team selected to build the Shuttle, and Phase D would cover manufacturing and test operations.

Phase A covered the period in which ILRV was studied, but in May 1970, NASA contracted two teams, led by North American Aviation and McDonnell Douglas, respectively, to complete Phase B definition studies. Initially, this embraced a common concept in which an Orbiter would ride on the back of a fly-back booster, each carrying a crew. Launched vertically, the booster would carry the Orbiter part of the way into space, release the Orbiter and fly back to a conventional landing using turbofan engines. The Orbiter would continue on into space on the power of two main rocket motors, returning to fly back down to a controlled landing after a week or two in orbit.

Meanwhile, on June 4, 1970, NASA asked econometric analysis company Mathematica Incorporated to carry out a funding analysis of a launcher program operating with a reusable Shuttle against one relying exclusively on conventional, expendable rockets. The 1960s were a time of growth in the satellite industry, and there was a burgeoning traffic model upon which to base projections. The baseline comparison envisaged a program of 12 years beginning in 1979. For that, Mathematica Incorporated imported a number of satellites launched between 1964 and 1969, an average of 61 a year It ran the model for separate space programs based around reusable versus expendable launch systems and came to an astonishing conclusion.

It found that if a reusable Shuttle program was solely responsible for a manifest involving 736 flights over 12 years, it would save considerable money over one relying on expendable rockets alone, even when the development cost of the Shuttle was factored in. This gave credence to the argument for a reusable Shuttle, and it was with these figures that NASA began lobbying both the White House and Congress for support and for approval, bringing the USAF along as a very influential ally; it did, after all, like the idea of saving on launch costs through a system paid for by NASA!

A radical rethink

Two very important conclusions from the Mathematica report appear not to have been fully prioritized: in the overriding point of the cost savings in a Shuttle-based system derived from (1) reuse, refurbishment and updating of satellites in space or (2) after being returned for upgrading and relaunch. The Shuttle, combined with a Space Tug as a "fetch" vehicle, could revolutionize low-cost access to space, said Mathematica. But, if there was no Tug, the economic case would be bleak. The second factor was that supporting a 12-year program involving 736 satellite launches would require funds that NASA would probably never receive in the sample period of 1979–1990.

One very big assumption NASA never seriously questioned was this optimistic launch rate of 736 satellites. NASA used data from Pan American Airlines to plan on flying the Shuttle at least once every two weeks – 26 flights a year. However, this assumed that this vehicle, which would push technology to the very limit of 1970s' capabilities, without a precedent to gain knowledge from or parallel against which to compare it, would routinely fly with such frequency. These engineering-orientated issues began to emerge in the second half of 1970, where a fully reusable, two-stage Shuttle was becoming increasingly implausible. Elegant as they were in concept, both Booster and Orbiter were very big, each carrying a full propellant load for their respective operations.

Above left: **The Titan Intercontinental Ballistic Missile (ICBM) was the favored candidate for launching the Dyna-Soar boost-glider into space, and flights were scheduled for the mid-1960s. (USAF)**

Above right: **All but one of the pilots selected to fly the Dyna-Soar were from the USAF and included (from left to right) Albert H. Crews, Henry C. Gordon, James W. Wood, Milton O. Thompson (NASA), Russell L. Rogers, and William J. Knight. (USAF)**

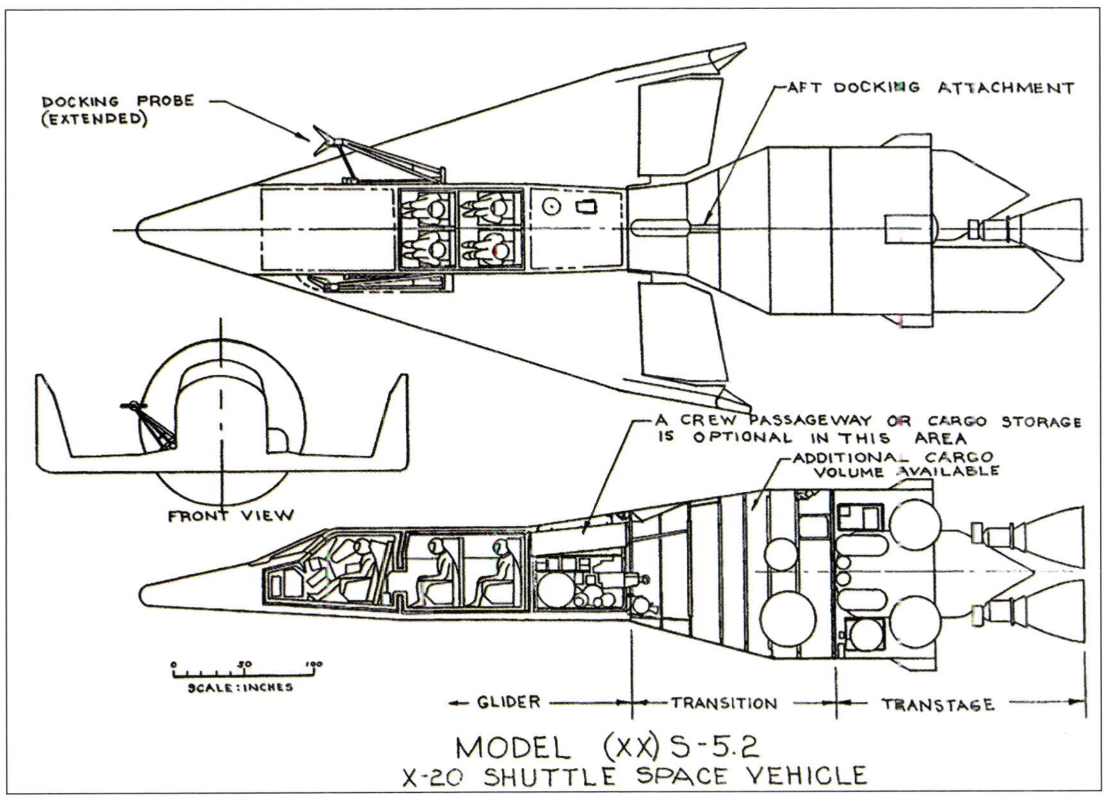

The USAF envisaged a wide range of missions for the Dyna-Soar/X-20 and studied extended versions capable of carrying several crew members, with a Transtage booster pack placing the spacecraft in low Earth orbit. (Boeing)

Moreover, when presented with the full estimated US$10bn development cost of the fully reusable, Booster/Orbiter system, Congress balked and demanded that NASA cut that in half. During the first half of 1971, the Phase B studies shifted toward a progressive slimming exercise – something had to be done to get the size and weight down while retaining the payload capability demanded by the USAF, without whose support there may not be approval.

First off, NASA asked the Phase B contractors to study several different concepts. The first looked at placing the liquid hydrogen fuel in jettisonable tanks on the top of the Orbiter's wings. Liquid hydrogen has 75% of the volume and less than 20% of the weight of liquid oxygen so removing them from inside the Orbiter reduced its size and weight considerably. Reducing the mass coming back flown through the atmosphere also helped to lower heat stress on re-entry.

The first option was to dispense with the massive fly-back booster and its crew, with 12 rocket engines in the tail and up to four turbofan engines. Instead, the Orbiter would be launched by a new, efficient and recoverable liquid-rocket booster or an existing adapted booster such as the first stage of a Saturn V. There was an attraction in using existing rocket stages to double as Shuttle boosters, but that rather belied the whole point of the Shuttle – that it be reusable – and there were few advocates for successfully recovering a very expensive Saturn V stage from a salt-water splashdown!

A further, more logical refinement proposed removing the LOX tank from the Orbiter as well and placing both that and the liquid hydrogen in a huge external tank, to the side of which the Orbiter would be attached. The tank containing the two separate propellants would be mounted to the top of the booster. It would provide propulsion for the Orbiter from the time it separated from the booster until it got into orbit, when it would be released. Further, by moving all the Orbiter propellant into an external tank, the size of the Orbiter was halved, dramatically cutting its weight to a minimum but still retaining the full payload capability.

So far, all options envisaged an Orbiter launched by a booster of some kind in sequence, with the Orbiter engines only ignited at altitude. By the end of 1971, there were several other deep cost-cutting options, whereby the Orbiter's rocket motors would ignite on the launchpad along with two boosters, placed either side of the external tank in a parallel-burn concept. These twin boosters could be either liquid propellant or solid propellant, the latter having the lowest impact on launch costs. Both types of boosters would be reusable after splashing down in the Atlantic Ocean after separating from the external tank. It all came down to how much it would cost to develop and how much it would cost to launch. All of these measures had a direct impact on whether it could get approval – first from the White House to get on the overall national annual budget request, and then from both houses of Congress, which in the United States is the only body capable of approving government projects and programs. The original fully reusable, fly-back booster concept would have a launch cost of around US$5m. The parallel-launch concept, with solid rocket boosters, would cost around $5.15bn to develop, almost half the original estimate rejected by Congress. And each launch would cost $10m – twice as much as the original concept.

Winning support

Over the Phase B period, the USAF grew to accept the Shuttle as a potential replacement for the old, expendable launch vehicles it had been using and gave the program its backing. But it drove a hard bargain and pushed NASA into decisions over configuration and, as a consequence, issues governing thermal protection. That came about because the USAF wanted to baseline the payload capability at 18,144kg (40,000lb) to a polar orbit, which required the Shuttle to launch from a completely new launchpad at Vandenberg Air Force Base (AFB) north of Los Angeles, California. The polar orbit flight had to fly south over the water for safety reasons and could not do that from Cape Canaveral owing to the South American landmass.

The USAF had a highly classified spy satellite that it wanted to launch by Shuttle, and it also wanted to refuel that type of heavy spy satellite in orbit to extend its operational life. Polar orbits

In 1961, when United States Secretary of Defense Robert McNamara lost interest in Dyna-Soar, it was de-rated to an experimental type, the X-20, and later cancelled outright in 1963. NASA, meanwhile, explored various shapes for re-entry vehicles in the Lifting Body program, including (left to right), the X-24A, the M2-F3 and the HL-10. (NASA)

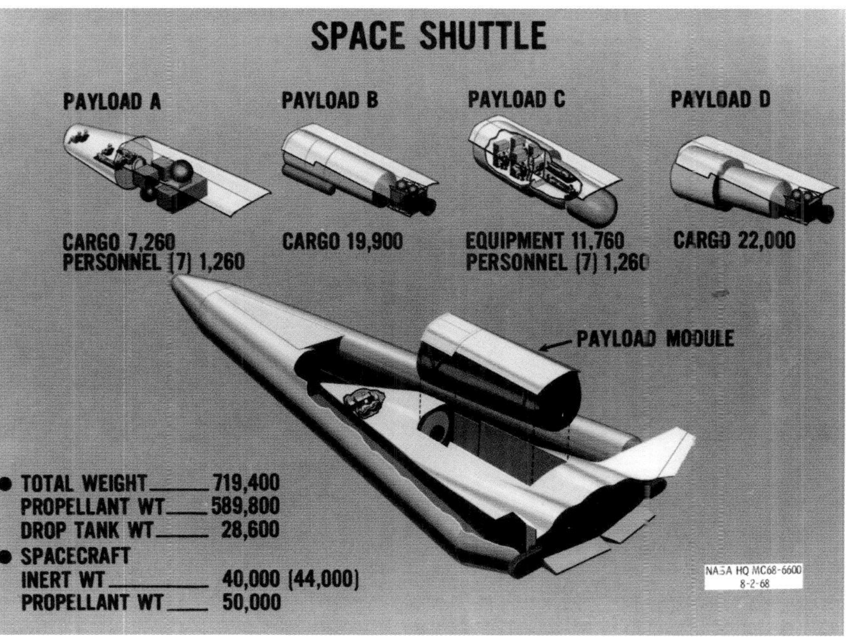

Above: In October 1968, NASA asked the aerospace industry to come up with a combined launch vehicle and reusable spacecraft, a so-called Integral Launch and Re-entry Vehicle (ILRV) to replace existing expendable rockets and single-use capsules. (NASA)

Right: Seen here depicting a range of potential mission models, studies had been underway for some time, and the term "Space Shuttle" had been coined by NASA manned flight boss George Mueller in a presentation to the British Interplanetary Society. (NASA)

allow the Earth to rotate on its axis while the satellite remains fixed with respect to the center of the planet. In such an orbit, when timed correctly, a spy satellite can pass over the same longitude once a day, maintaining a continuous surveillance over a specific set of locations. This had two negative consequences. The USAF wanted to launch from Vandenberg, deploy the satellite and land within one revolution of the planet. The spin rate of the Earth means that in one orbital period of 90 minutes, the Earth has rotated 2,400km (1,500 miles) in an anti-clockwise direction. To get back down to a landing site at the same longitude, in this case Edwards AFB, California, the Shuttle would have to fly that distance to the left of its south-moving ground track. It is for this reason that optimum launch locations for satellites close to the equator or at orbital inclinations up to about 50 degrees launch out of Cape Canaveral to gain advantage from the Earth's rotation.

And that brings yet further problems, specifically adding a payload requirement to the Shuttle. Because polar-orbit launches do not have the advantage of getting a boost from the Earth's spin, less payload can be carried. Hence, the 18,144kg (40,000lb) polar-orbit payload is comparable to a 19,165kg (65,000lb) payload when launching out of Cape Canaveral. However, this might be thought of as beneficial for launching space station modules and general satellite traffic. In fact, the demand from the USAF pushed the payload mass much higher than NASA felt it required, at least when the Phase A studies took place. In time, NASA did come to make the most use of that capability, but the capacious size of the payload bay, 18.2m (60ft) by 4.5m (15ft), was ultimately driven by the need to carry the USAF spy satellite.

By far the most challenging issue to overcome was the cross-range requirement, another USAF stipulation, which would allow for the substantial maneuver necessary during descent to correct for the effect of the Earth's spin. That demanded a delta wing to get the lift to make those maneuvers whenever necessary to get back to the landing site close to where it had been launched from. By using a delta wing planform and accommodating that cross-range, the thermal protection was pushed to higher requirements than would have been necessary with a smaller, straight-sing Orbiter originally proposed by NASA.

As NASA halved the development cost, thus doubling the price of each flight, the USAF testified to use of the Shuttle, and this did a lot to get it approved by the White House. In January 1972, President Nixon gave his formal blessing and presented it to Congress, but largely as a stimulus to jobs and

Above left: One potential concept involved a piggyback configuration of booster and Orbiter, the former returning to a piloted landing after pushing the Orbiter part way out of the atmosphere and into orbit. (NASA)

Above right: Scale models of the piggyback shuttlecraft and a small airliner provided a clear size comparison between a commercial aircraft and the orbital aerospace vehicle. (NASA)

industry. Three months later, NASA asked the industry for their proposals, and the contract went to North American Rockwell in July 1972, five months before the last Apollo Moon landing. The contract to build the Orbiter's rocket motors had gone to Rocketdyne in July 1971, the technological demands making it the long-lead pacing item.

Congress backed the revised design, and the decision was made to adopt solid rocket boosters in a parallel-burn sequence, with initial flights into space being scheduled for 1979. But there were still cost savings and design tweaks to be made. In the event that something went wrong after launch and before the boosters were jettisoned, a small solid rocket motor had been placed on top of each wing either side of the tail to carry the Orbiter to safety, but these were deleted to save weight. Furthermore, the two turbofan engines originally envisaged to help the Orbiter fly safely to a runway were removed in the full expectation that they would not be needed.

It was expected that the Orbiter would be moved around on top of a specially converted Boeing 747, and this would be the way it would first feel the

An optional alternative fell back on a delta-wing version of the X-15 and married that to a fly-back booster. (Boeing)

atmosphere as a glider, "flying" off the top of the Jumbo Jet for test flights to evaluate its aerodynamic qualities before committing it to a launch and orbital operations. There were many technical complexities with this, not least being the development of the thermal protection system made all the more demanding by virtue of the type of trajectory it could be called upon to fly on returning from orbit. Operating such a large and complex machine required a sophisticated computerized flight control system and a diagnostic capacity that no other aerospace vehicle boasted.

Moreover, continuous communications would be essential, rather than intermittent coverage as it passed from station to station around the globe. For that, a new and highly sophisticated tracking and data relay satellite system (TDRS) was planned to support round-the-clock contact. This in itself would revolutionize the way information would flow, not only to and from the Shuttle but also to a multitude of satellites and spacecraft in low, medium and high orbits.

In 1973, As the Apollo lunar landings came to an end, NASA launched the Skylab space station, itself an outgrowth of redundant hardware from the Moon program, with three teams of three astronauts raising the all-time space endurance record to more than three months in 1974. In 1975, the very last Apollo mission conducted a rendezvous and docking handshake with Soviet cosmonauts in a joint mission of collaboration and a unique display of détente. For NASA, the second half of the 1970s were dominated by the reusable Shuttle.

With contracts in place for four Orbiters – one to fly off the back of a Boeing 747 for aerodynamic tests and three for space missions – work intensified on adapting the two Saturn V launch pads at Launch Complex 39 (LC-39). On the other side of the continent, preparations got under way for building a Shuttle pad at Vandenberg AFB, from where the USAF would launch its classified missions. All the while, NASA intensified its efforts to recruit customers for the Shuttle, assembling a manifest of commercial payloads from governments across the globe.

This was the dawn of the first commercial space age, with many countries wanting their own communications, TV, weather, and resource-monitoring satellites. Rather than paying for an expendable rocket to place them in orbit, NASA offered foreign customers budget-price Shuttle-launched slots, several carried up into space on a single launch. It looked for all the world as though Mathematica had got it right; the Shuttle really would stimulate a renaissance in space activity.

Left: The enormous size of such a concept is displayed alongside this model of a B-52. (Boeing)

Below: While NASA was still looking at feasibility studies, several companies put forward their own proposals, including this Boeing-Lockheed concept. (NASA)

At the beginning of Phase B studies in mid-1970, NASA asked for proposals for Orbiters with high cross-range (delta wing) and low cross-range (straight wing) capability, defined as the range capability flying to left or right of the ground track during re-entry. This is Grumman's concept for the low cross-range Orbiter. (Grumman)

Above left: An artist envisages staging with the low cross-range concept proposed by the Manned Space Center, Houston, Texas. (NASA)

Above right: At the start of Phase B studies for a reusable shuttlecraft, NASA was managing study contracts with North American Rockwell (NAR) and McDonnell Douglas for a 12-person space station launched by two stages of a Saturn V. This depiction shows the NAR concept with shuttlecraft docking ports. (NAR)

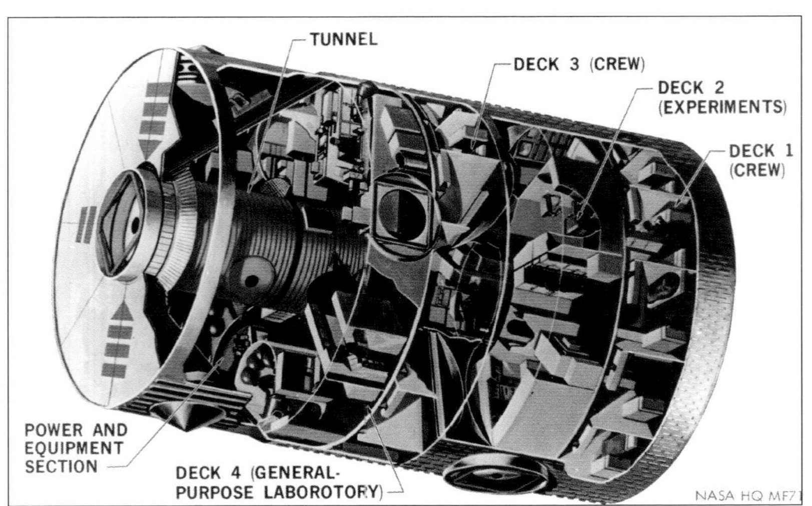

The McDonnell Douglas space station Phase B contract produced this design for its concept, again with multiple docking ports for a shuttlecraft, which was, in 1970, the prime reason for wanting to build the reusable vehicle. (MDAC)

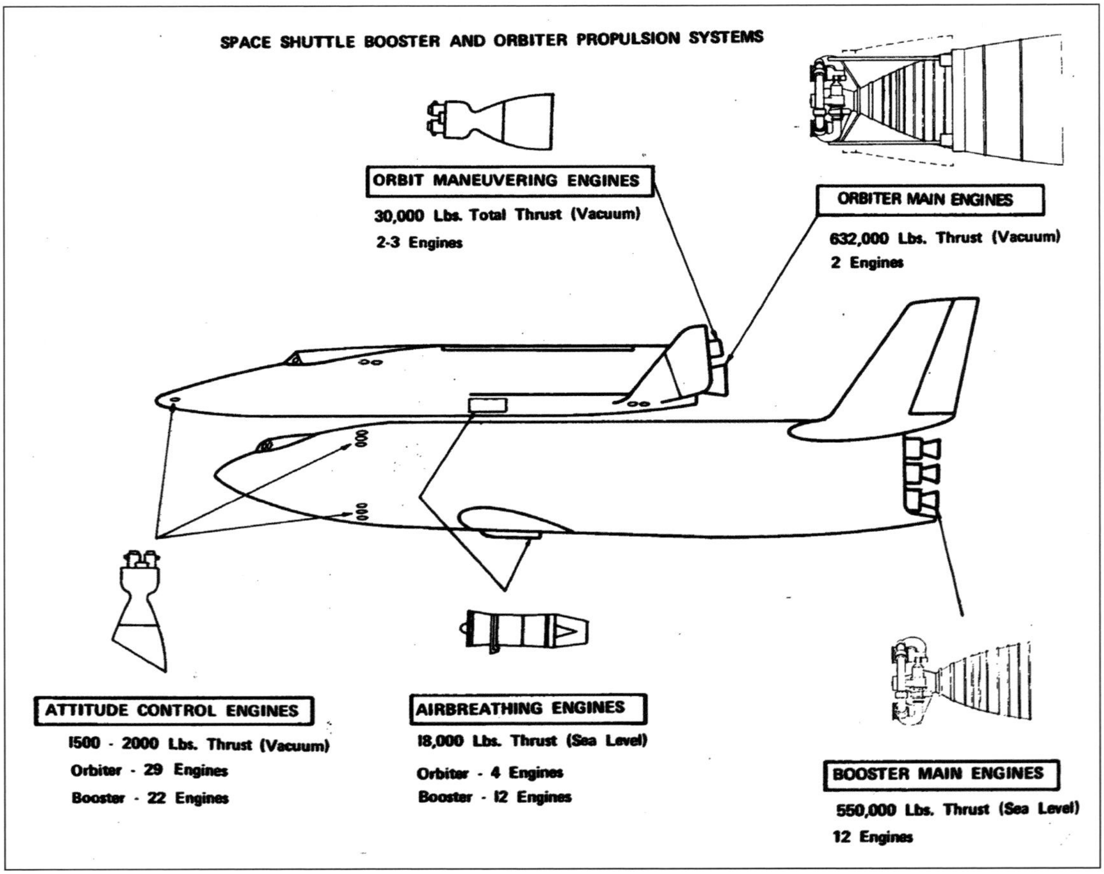

This is the baseline Phase B starting point selected by NASA for a fully reusable two-stage Booster/Orbiter combination for contractors to work to and propose their respective concepts. Note the multiple rocket motors for each vehicle, air-breathing engines for independent ferrying around the country, and attitude control thrusters. (NASA)

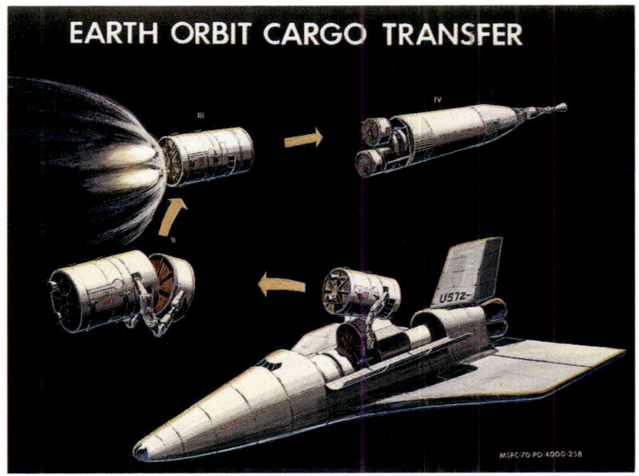

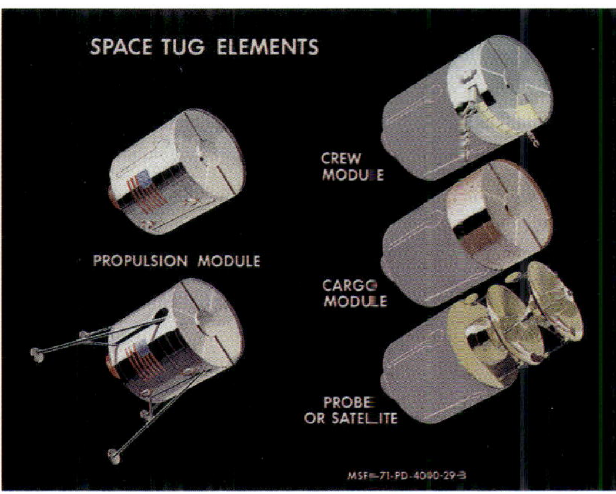

Above left: During Phase B studies, the station was deferred until after the Shuttle and a wider range of capabilities were imported to broaden the economic viability of this reusable vehicle, particularly using a Space Tug to move heavy payloads around to orbits the Shuttle could not reach. (NASA)

Above right: The Space Tug was envisaged with manned and unmanned versions according to requirements. Quite quickly, the Tug also fell by the wayside, as it was a cost the NASA budget could not support. (NASA)

The fully reusable Shuttle would be launched from converted launch pads at LC-39, Kennedy Space Center, which had been built for the Saturn V Moon rockets. (NAR)

Above: Powered by 12 rocket motors of the type used on the Orbiter, the booster propels the shuttlecraft into space before separating and returning to a piloted landing on a conventional runway. (NAR)

Left: With a delta-wing configuration, the high cross-range Orbiter separates from the booster on its way into space. (NAR)

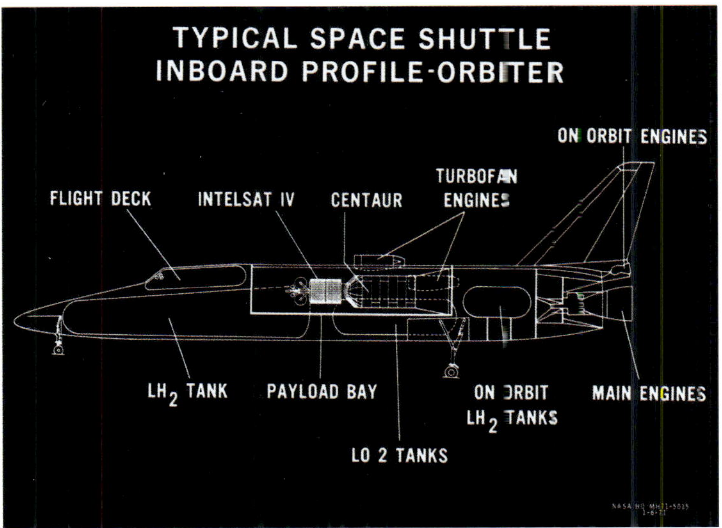

Right: A fully reusable Orbiter, with integral liquid hydrogen and liquid oxygen propellant tanks for the two or three main engines, together with an Intelsat IV communications satellite attached to a cryogenic Centaur stage for propelling it to geosynchronous orbit, where it will operate. Note that, at this stage, turbofan engines could be attached for ferrying the Orbiter. (NAR)

Below: The low cross-range Orbiter re-enters the Earth's atmosphere. (NAR)

Left: The McDonnell Douglas initial Phase B design concept had a noticeably pointed forward fuselage and a booster with twin vertical tails. (MDAC)

Below: High and low cross-range Orbiters from North American Rockwell and McDonnell Douglas Aircraft Corporation (MDAC) compared to the Saturn V in scale. Note the folded wings for launch of the MDAC design and that company's parallel-burn, twin booster concept for its high cross-range Orbiter. (NASA)

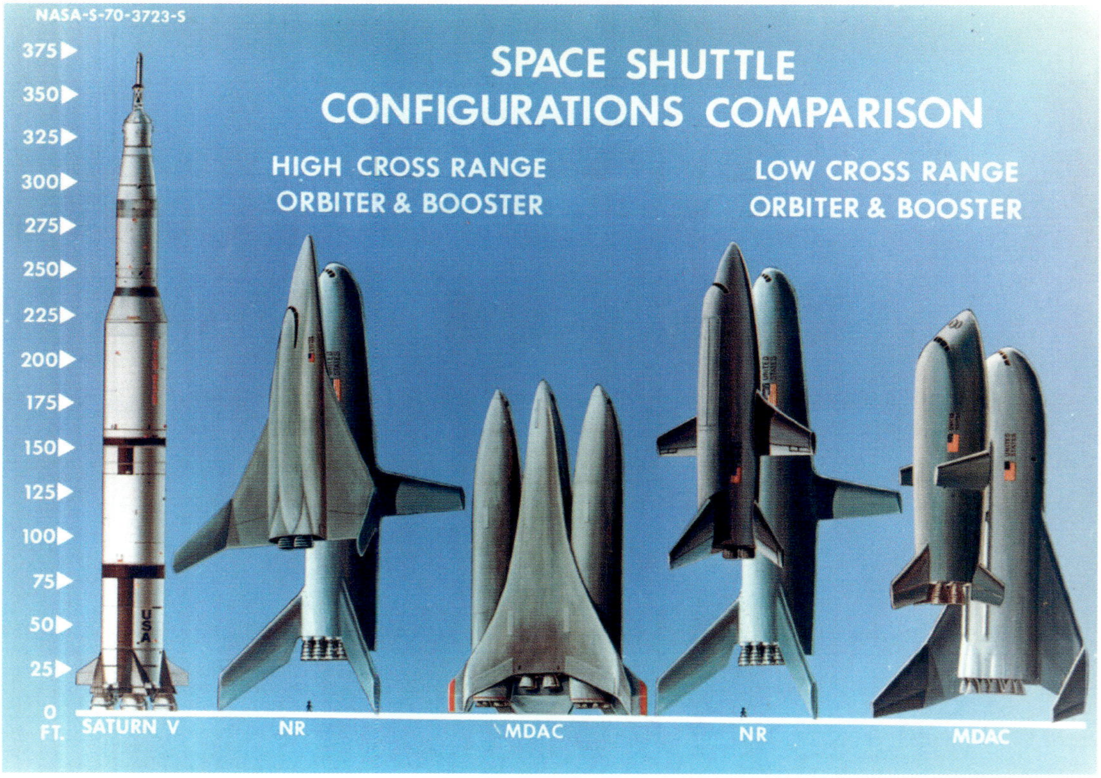

Extensive wind tunnel tests were conducted to evaluate the aerodynamic properties of different Orbiter configurations using extrapolated data from unmanned tests and manned spacecraft. (NASA)

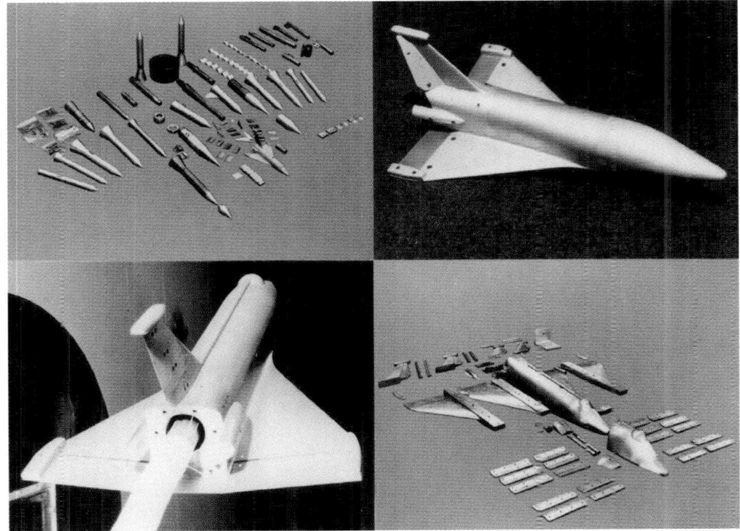

In attempts to reduce the weight and length of the Orbiter and cut costs, it was proposed that the liquid hydrogen could be carried in over-wing tanks, which would be jettisoned after reaching orbit. (NAR)

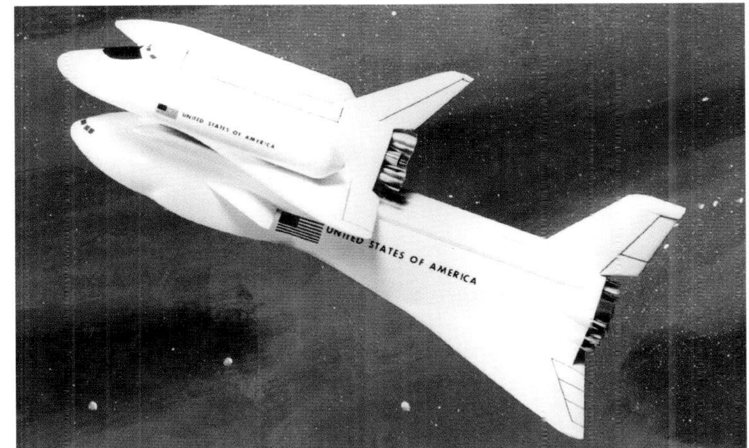

All Phase B contenders incorporated external hydrogen tanks into their proposals but retained the piloted, fly-back boosters. (NASA)

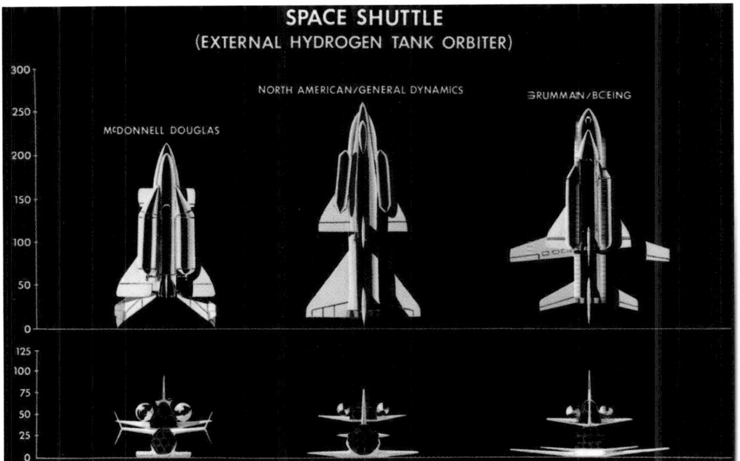

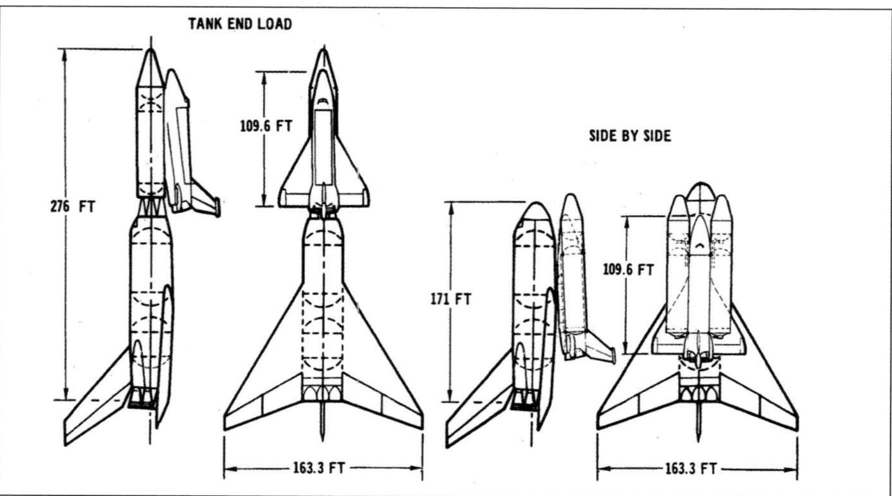

In further developments to reduce size and save weight, McDonnell Douglas proposed a reusable fly-back booster powered by F-1 engines, developed for the Saturn V, with either all the Orbiter propellant and its rocket motors in a single expendable eternal tank (left), or with over-wing tanks feeding engines in the base of the Orbiter (right). (MDAC)

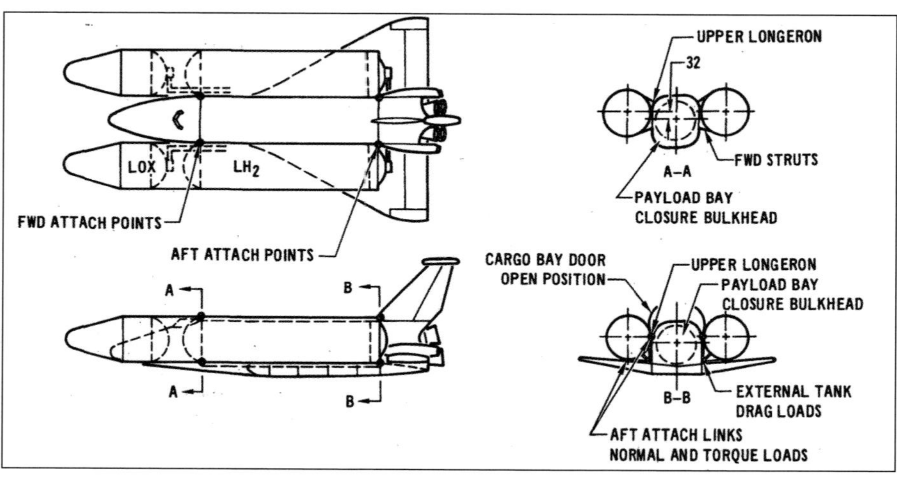

The McDonnell Douglas proposal for Orbiter over-wing propellant tanks shows how they would be attached to the fuselage. (MDAC)

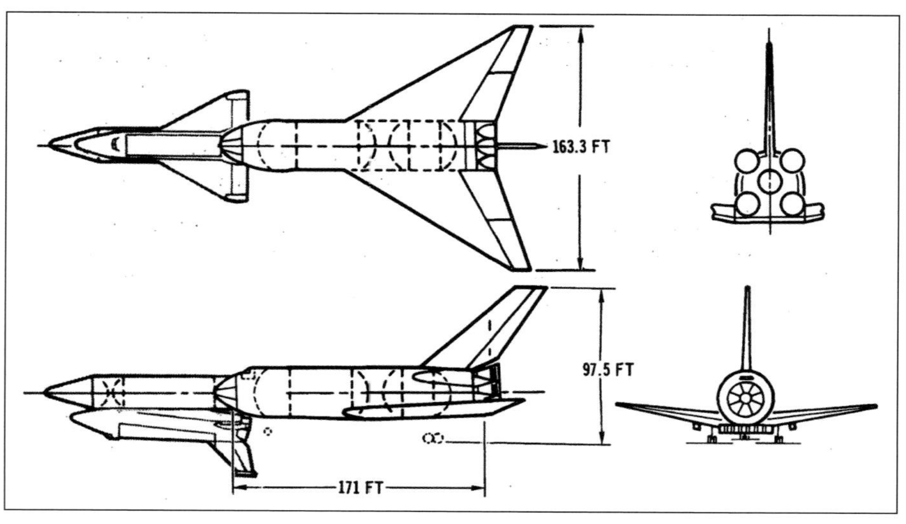

The alternative configuration from McDonnell Douglas, with an external tank (ET) attached to the winged fly-back booster. (MDAC)

When NASA started looking around for a reusable booster of conventional configuration, McDonnell Douglas examined a pressure-fed liquid propellant system powered by liquid oxygen and propane, the stage being recovered and used again. (MDAC)

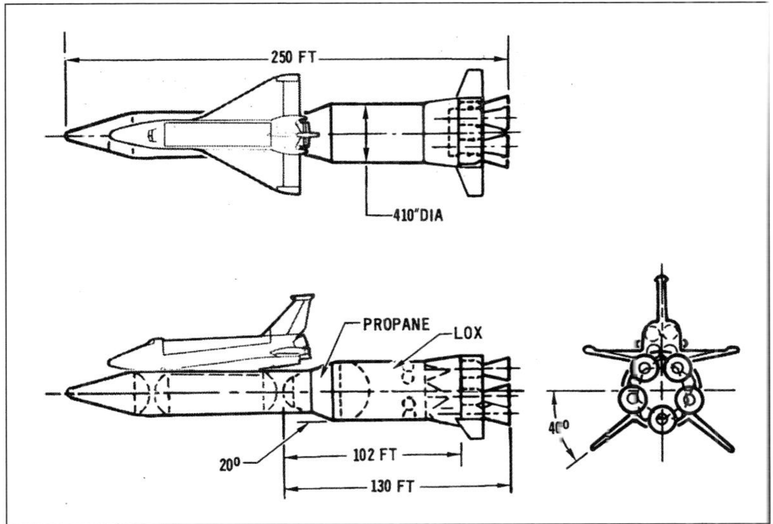

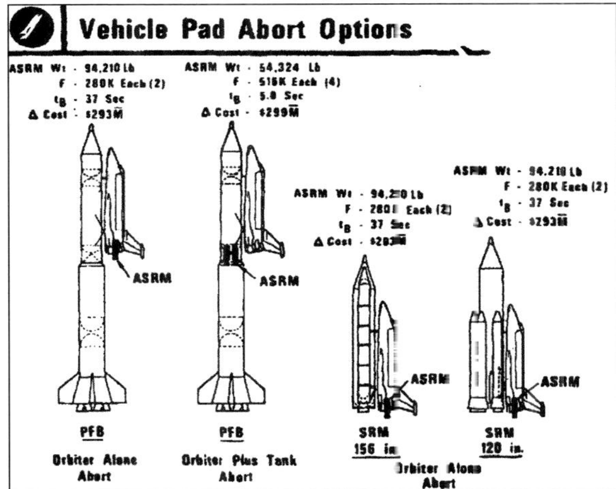

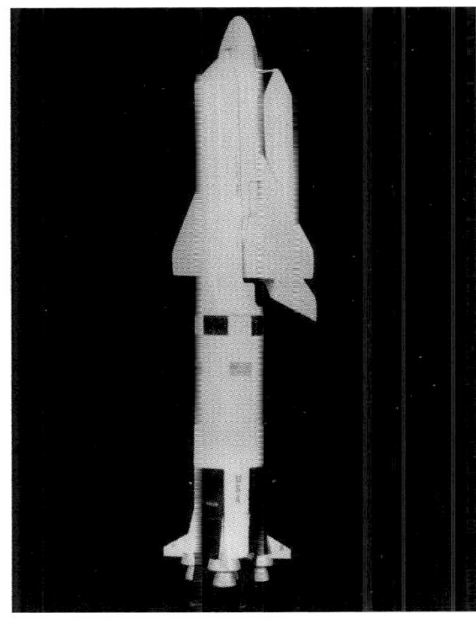

Above left: Analyses showed that instead of a series-burn, two-stage configuration, savings could be made with a parallel-burn concept in which booster and Orbiter engines all ignited on the pad. Here, models of two parallel-burn configurations are shown with liquid propellant boosters (left) and solid propellant (center) with a series-burn liquid propellant booster. (NASA)

Above right: A pad abort chart demonstrates crew safety on the conceptual contenders: two pressure-fed liquid boosters (PFB) and two solid-rocket motors (SRMs). The Abort Solid Rocket Motor (ASRM) devices were solid propellant rocket motors for lifting the Orbiter free of a booster running amok during ascent. In still further cost-cutting, these too were removed. (NAR)

Right: During the final flurry of configuration bids, Boeing proposed use of the entire S-IC first stage of the Saturn V rocket, which was, in 1971, still carrying men to the Moon and would be used in 1973 to launch the Skylab space station. (NASA)

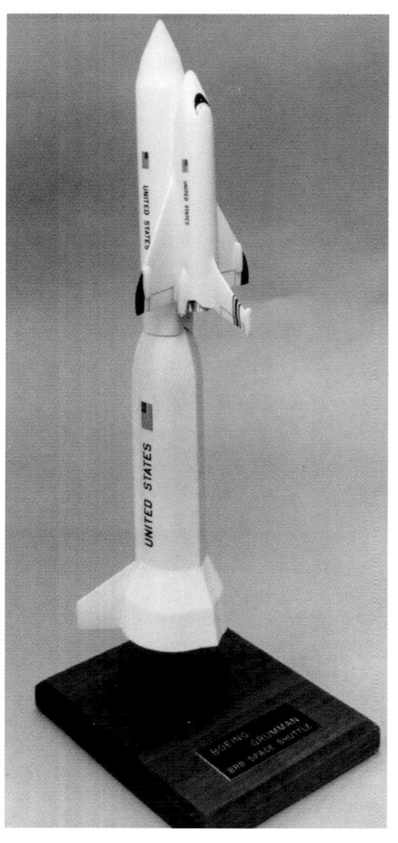

Above: Wind tunnel tests were conducted on the parallel-burn configuration to display shock waves at various altitudes and levels of dynamic pressure. (NASA)

Left: In a further option for the use of the Saturn V first stage, Boeing teamed up with Grumman and proposed a new liquid propellant booster, more efficient than the S-IC. (NASA)

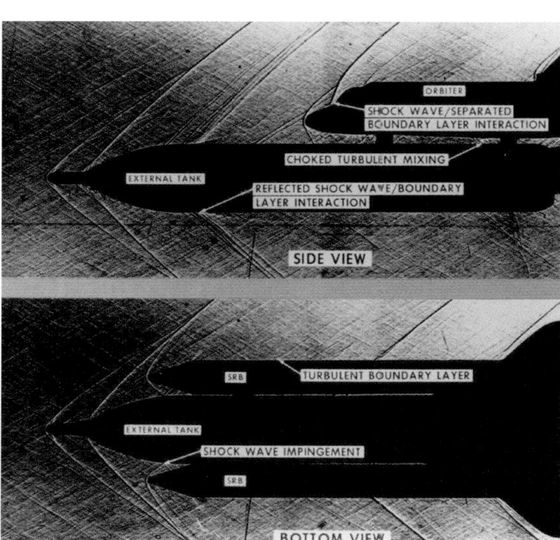

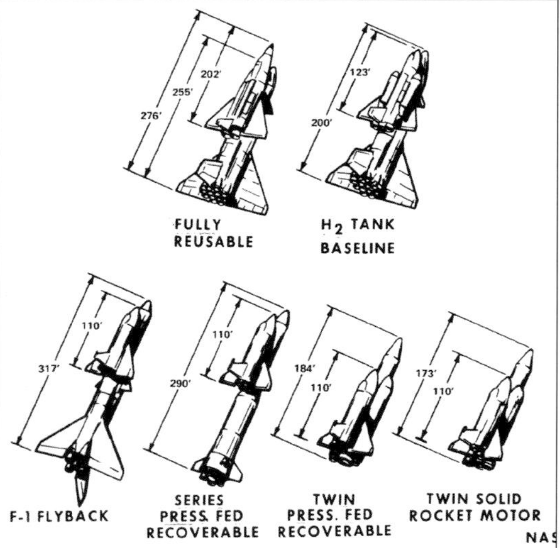

Above left: Schlieren imaging of shock waves on the full Shuttle assembly shows the effects of subtle changes to the exterior shape of the Orbiter. (NASA)

Above right: The evolution of the Shuttle configuration from the fully reusable system at the top left to the parallel-burn, solid rocket booster configuration at the bottom right. (NASA)

The configuration of NAR's Shuttle. Note the ASRMs, one of which is seen on the starboard aft inboard wing section. Also note the turbofan engines, proposed as an independent means of flying it back from landing site to launch site together with its fuel tank. The forward section of the SRBs have blow-out ports for thrust termination to shut them down in the event of a malfunction. This also shows the ET nose cover over an abort motor to decelerate the tank out of orbit. All of these features were eliminated. (NASA)

As envisaged by contract winner NAR, the Shuttle heads for space, now missing the SRB blow-out ports and the ET retro-rocket nose cap. (NAR)

Chapter 2

Shuttle Design and Assembly

Although technically a spacecraft, the Shuttle Orbiter looks like a conventional airplane. That shape is dictated by the requirements of stable and controlled flight back down through the atmosphere to execute an unpowered landing at a predesignated runway. It could have any shape to achieve orbit – within reason. But the essential systems could not be done without a pressurized crew compartment, a large payload bay to carry the cargo, provision for three rocket engines at the rear, a delta-wing with elevons, and a dorsal tail for directional control.

In addition, the lower section of the fuselage beneath the payload bay contained the oxygen and nitrogen tanks for providing a "shirt-sleeve atmosphere" similar to that at the surface of the Earth, and oxygen and hydrogen tanks for fuel cells providing electrical power and cooling. Either side of the vertical tail, two pods contained propellant and rocket motors for maneuvering in orbit and for deorbiting the vehicle at the end of its mission. The Orbiter had a tricycle undercarriage, and electronics and computers forward of the crew compartment. Thrusters for attitude control were situated in the nose and on the exterior fuselage pods.

In several aspects, the structural design and layout were the ultimate in simplicity. The fuselage was in three distinct assemblies: the forward section, the mid-fuselage and the aft fuselage. The forward

In every essential aspect, the Shuttle Orbiter followed the pattern of a conservatively designed aircraft using innovative technology to achieve orbital flight and effect an unpowered controlled re-entry and landing. (NASA)

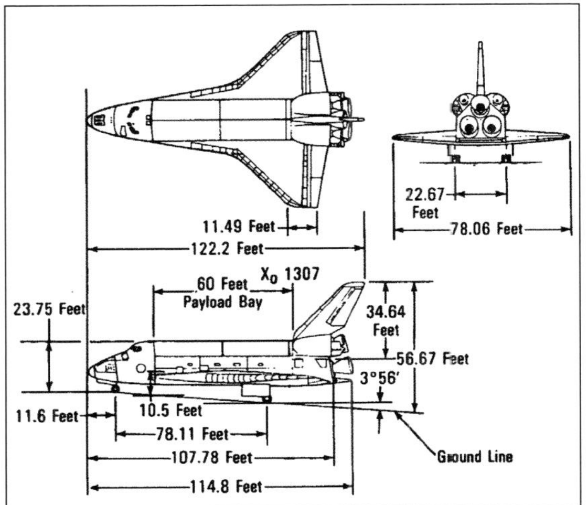

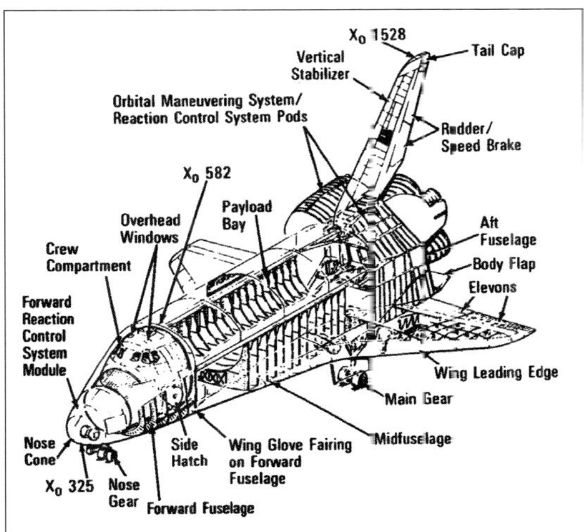

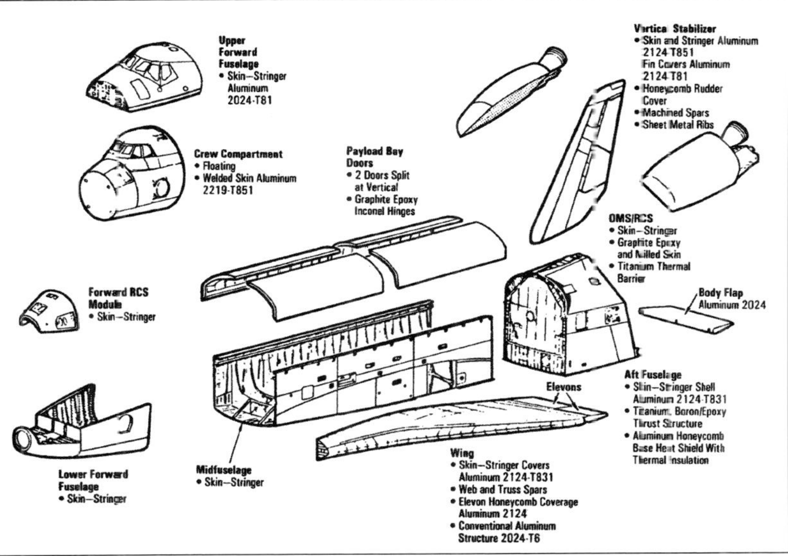

Above left: The definitive Orbiter was less than half the length of the initial concept, in which propellants for its main rocket motors had been contained within the fuselage. (NAR)

Above right: The basic Orbiter layout, with the disposition of rocket motors, attitude control thrusters and aerodynamic control surfaces. (NAR)

Right: The primary elements, most of which were contracted out to other aerospace manufacturers. The crew compartment was a separate structure within the forward fuselage within which it was encased. (NAR)

section included the pressurized crew compartment and the unpressurized nose, which contained the nose wheel leg and thrusters for attitude control. Of aluminum alloy, welded construction and attached to the mid-fuselage at four places, the forward fuselage was of skin-stringer construction with vertical frames riveted to stringer panels. It contained a 101.6cm (40in)-diameter circular side hatch for access while on the ground.

With a volume of 65.6m³ (2,325ft³) the pressurized area was divided into three levels, the mid-deck area accessed directly from the side hatch and containing several seats for the crew during launch and descent. Below was the equipment bay, which contained both vital systems that the crew might need access to and water tanks. The flight deck was above the mid-deck, where seats were provided for two pilots and two passengers. The left front seat was for the commander, and the right seat for the pilot. The other seats on the two levels were for mission or payload specialists. The interior was pressurized with a mix of 80% nitrogen and 20% oxygen at 101kPa (14.7lb/in²).

The crew cabin had 11 windows, six wrapped around the forward area for the pilots with two in the rear bulkhead looking back into the payload bay, two on top for looking directly up and one in the side hatch. The forward-facing windows were the thickest ever produced and consisted of three separate panels. The innermost panel was designed to withstand atmospheric pressure, the middle panel formed an optically pure shock screen, and the outer panel provided heat insulation and further impact protection. The inner and outer panes were each 1.5cm (0.6in) thick.

With a length of 18.29m (60ft), a width of about 5.18m (17ft) and a height of 3.92m (13ft), the mid-fuselage weighed 6,124kg (13,500lb) and was assembled from 12 main, vertical frames, each strengthened by boron/aluminum trusses reinforced with skin and longerons. This assembly provided a spacious payload bay, boxed in at each end by bulkheads to the forward and aft fuselage, and to which were attached the wings supported on a carry-through box at the floor.

The top of the mid-fuselage/payload bay was closed out by two doors 18.29m (60ft) long and with a width of 3m (10ft) across the radius. Each door was assembled from five sections and attached to each side of the mid-fuselage by 13 hinges, of which eight were "floating" to enable thermal expansion or contraction. The two doors closed to form a box structure with the mid-fuselage and were locked together with 16 latches along the centerline and eight at each end. The payload bay was not pressurized, but when closed it formed a relatively airtight structure with little thermal conductivity.

The aft fuselage provided a structural base for the vertical tail, a capacious void for three Space Shuttle Main Engine (SSME) motors, plumbing for bringing propellant for those motors from the External Tank (ET), an aft body flap for pitch trim, and two Orbital Maneuvering System (OMS) pods each containing rocket motors and propellant. It was about 5.48m (18ft) in length, 6.7m (22ft) wide and 6.1m (20ft) high. It formed the aft closeout on the mid-fuselage/payload bay assembly with

Above left: **The forward fuselage was stressed to accept the pressurized crew compartment and is seen here with the mid-body assembly that will become the payload bay. (NAR)**

Above right: **The aft fuselage section is now attached, and the crew compartment has been lowered into position and fixed within the forward fuselage. (NAR)**

The crew compartment contained the star tracker well for the guidance and navigation system and the side hatch, which was the only access or egress point for the three decks. (NAR)

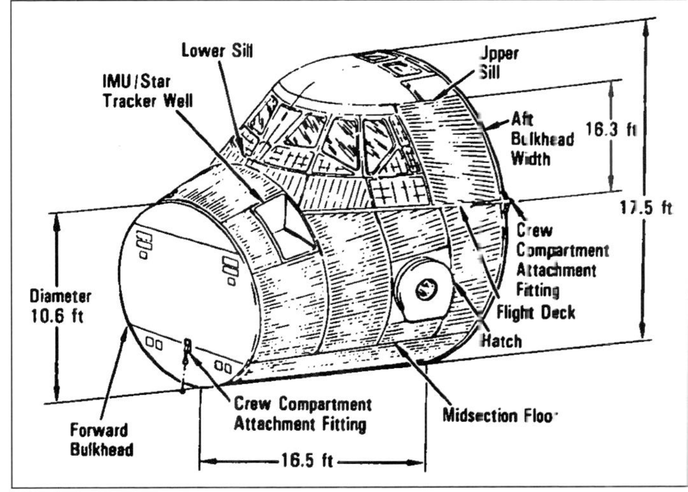

The flight deck provided seating for the two pilots, two seats for passengers behind them, two aft-facing windows looking into the payload bay, and two looking up above the Orbiter. (NAR)

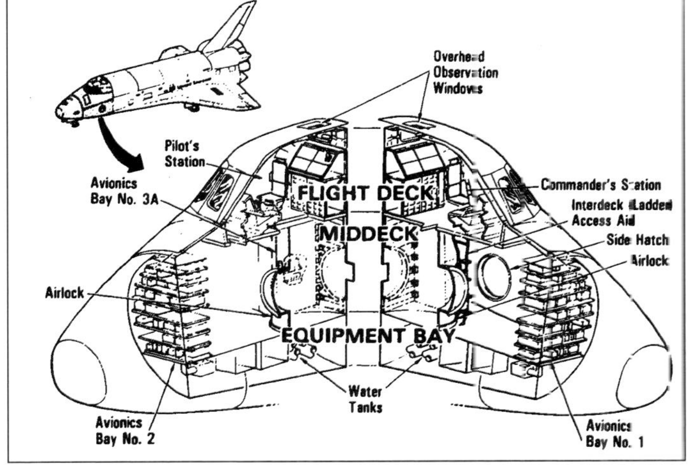

The mid-deck area provided space for lockers, sleep cubicles, food and clothing storage, and an access hatch to the lower deck containing systems and subsystems. (NAR)

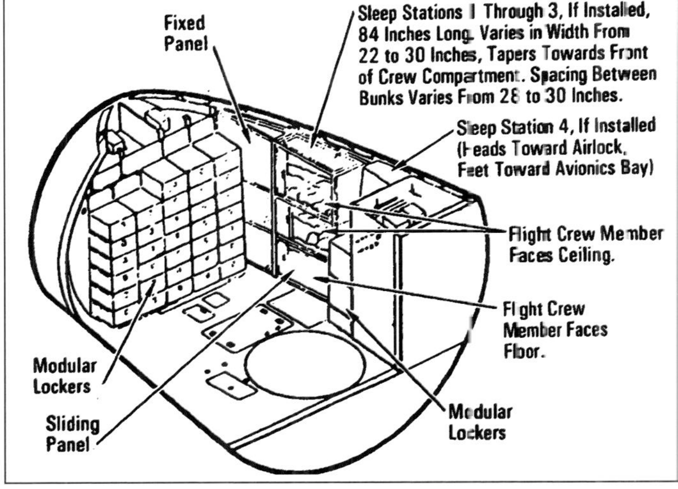

a bulkhead comprising machined and beaded sheet aluminum segments. The upper bulkhead was attached to the front spar of the vertical tail.

Directional stability in the atmosphere after re-entry was crucial to the safe return of the Orbiter. The vertical fin supported a split-rudder/brake assembly responsible for this and for decelerating the speed of the vehicle at various times as required during the descent to a safe landing. The fixed portion of the fin was attached to the structural surface of the aft fuselage at two locations. It consisted of aluminum ribs and spars, and integrally milled skins with a total height of 8m (26.25ft).

The rudder had a total height of 5m (16.6ft) and maximum width of 2.3m (7.5ft) at the base. Separate to the fin, but of essentially the same construction, the split-rudder was essentially of two halves co-located at the hinge line and attached by four hinges with hydraulic actuators capable of operating the rudder to left or right deflections of 27 degrees either side of the centerline, or of actuating the drive shafts in opposing directions to open the two halves to an included angle of 98.6 degrees.

The body flap was 6.4m (21ft) wide where it was attached to the fuselage, 2.2m (7.24ft) long and 4.56m (18.25ft) wide at its extreme end. It consisted of an aluminum structure attached to the fuselage and could move up 15.7 degrees and down 26.55 degrees. It was used only during re-entry and for flight through the lower atmosphere where it operated as a pitch trim and to regulate the trade between lift and drag. Being an unpowered glider for the entire duration of its free flight through the atmosphere, this was a driving consideration in the overall layout.

The wings were of relatively conventional design and construction, with which any aerospace engineer would be familiar. From the interface with the mid-body section of the fuselage, each had a forward wing boot, a torque box in the mid-section, and a wing/elevon interface box at the rear. The trailing edge supported inboard and outboard elevons, with each wing having separate main gear well assemblies. Each wing had a maximum chord of 18.3m (60ft) and a maximum thickness of 1.52m (5ft), with the forward box blending into the mid-fuselage position.

The wing was built around four main spars constructed of corrugated aluminum to minimize the transfer of thermal loads, the rear spar providing attachment for the elevon hinges. Accounting for

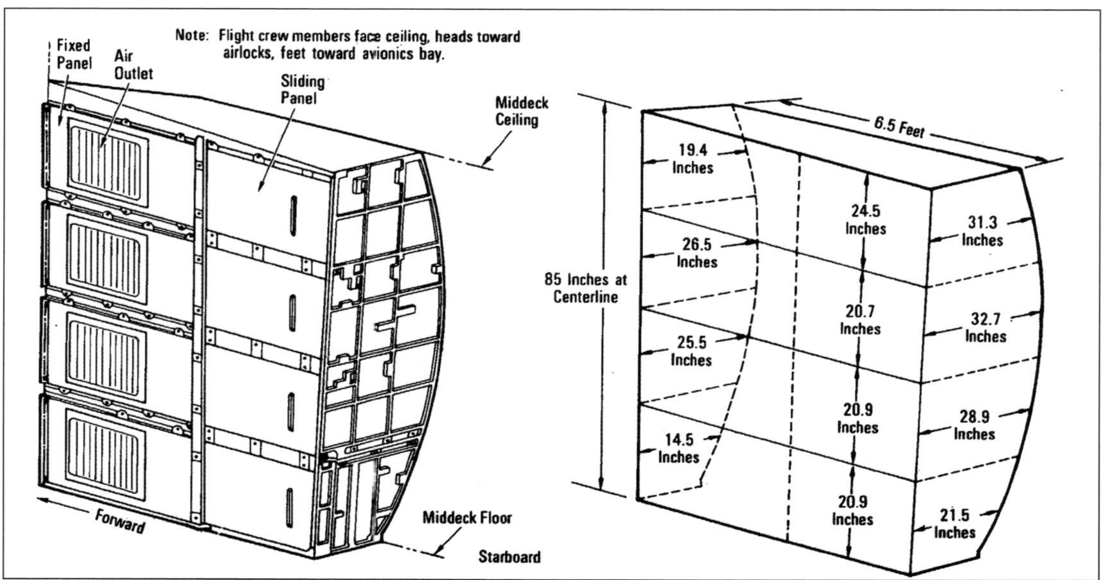

The configuration of sleep cubicles and lockers varied between Orbiter and were variously reconfigured during the 30 years of space operations. (NAR)

The forward avionics bay, which allowed access to the General Purpose Computers (GPCs) and electronic systems, was located behind the forward lockers and faced forward toward the nose. (NASA)

20% of the total wing area, the elevons covered the entire trailing edge, the inboard section with a spanwise length of 4.2m (13.8ft) and the outboard with a length of 3.8m (12.4ft). The elevons tapered in width from 2.6m (8.7ft) inboard of the inner elevon to 1.2m (3.9ft) at the extreme outer edge. Driven by hydraulic actuators, they could move 40 degrees up and 25 degrees down. Each elevon was fabricated from conventional aluminum rib beam construction and supported by three hinges. The main landing gear doors were located in the intermediate section of each wing and were 3.8m (12.6ft) long by 1.5m (5ft) wide.

Although the basic design of the Orbiter structure remained the same, over time great improvements and weight-saving changes were made to successive vehicles. Although only five flight-rated Orbiters were built and flown in space, their manufacture spanned a period of more than 12 years, and much was learned along the way. No two Orbiters were identical, and adaptations in construction, methods and practices greatly improved the performance of each vehicle, sadly only a little of which can be recorded here. Of particular note were changes involving the evolution of the thermal protection systems, which are explored below.

Orbiter operating systems

Thermal protection was crucial to the successful return of the Orbiter through the atmosphere at speeds of up to Mach 20. Previously, in all manned space vehicles, thermal protection was achieved through ablative materials designed to char and burn away as the temperature exceeded its phase change. Usually, an epoxy resin injected into an aluminum honeycomb matrix, this material was heavy and would have to be reapplied after every flight, which was why ballistic capsules were used only once as this was too expensive, time-consuming, and inefficient. However, as the Shuttle was designed to be reusable, it required a material that could remain intact on successive flights.

Upon re-entry, the maximum temperature of the Orbiter nose and wing leading edges would reach almost 1,648 degrees Celsius (3,000 degrees Fahrenheit). To protect the vehicle's structure against this, NASA selected a Reinforced Carbon-Carbon (RCC) material, a composite of carbon fiber and graphite. It possessed a high thermal shock but low coefficient of expansion, making it very brittle and vulnerable to damage, but, crucially, it could be used repeatedly on successive flights without significant refurbishment.

The nose cap, which protected an area devoid of thermal expansion or contraction, had a solid covering of RCC. The opposite was the case with the wing leading edges. If applied as a continuous

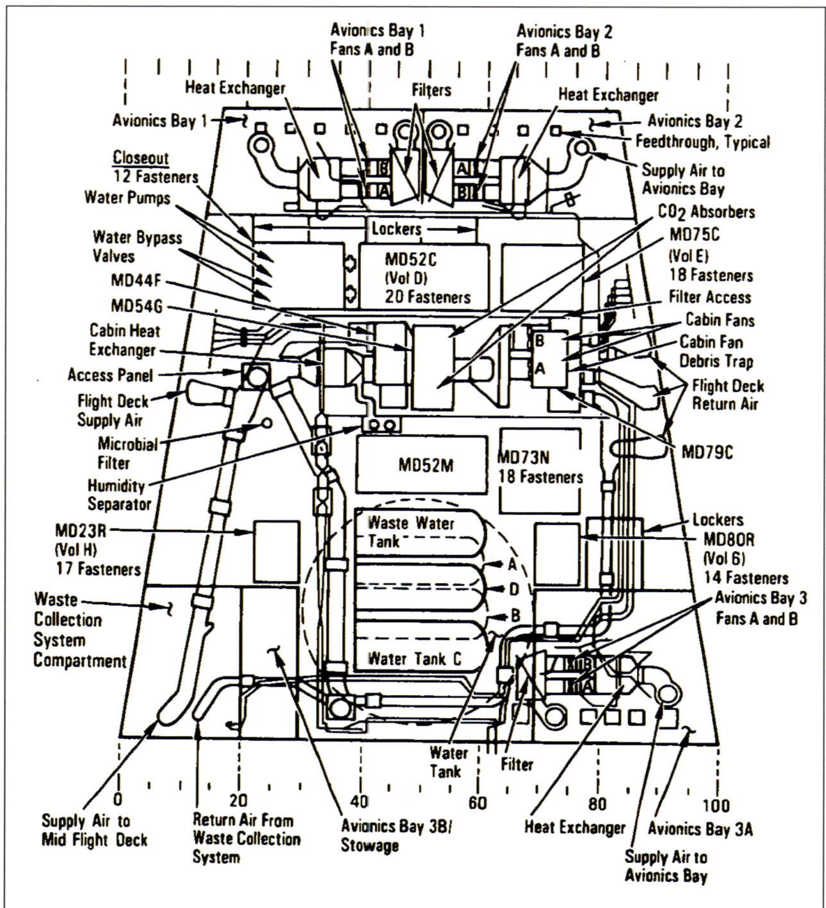

Left: The essential layout of the lower equipment bay, looking down below the mid-deck with the forward section at the top. The area was not habitable but was accessed through the floor of the mid-deck. (NAR)

Below left: A snaking line of technicians carry preassembled wiring looms and bundles of cables for installation in the pressurized compartment. (NAR)

Below right: The top of the forward fuselage is attached to the lower section, fully encapsulating the insulated and pressurized crew compartment. (NAR)

Subcontracted to General Dynamics in San Diego, California, the mid-fuselage section consisted of the keel for the Orbiter, attachment points for the wings, the payload bay doors, and various life support and electrical power systems. (NAR)

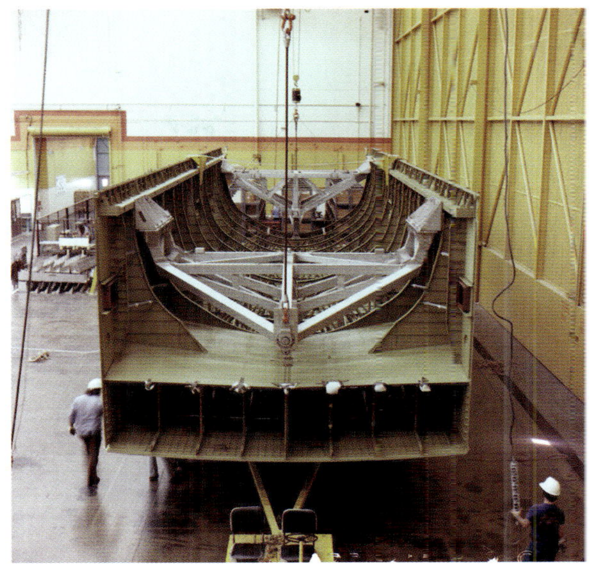

sheet, the flexing of the wing structure with heat soak and temperature changes during re-entry would cause the RCC to shatter. To avoid that, it was necessary to apply the RCC as a series of 22 panels attached to the forward spar, with expansion and contraction joints between each one. In this way, the wing structure could move and flex without cracking the RCC.

During re-entry and for much of the flight through the upper atmosphere, the Orbiter would be pitched up 27 degrees, exposing the underside to temperatures as high as 1,427 degrees Celsius (2,300 degrees Fahrenheit). These surfaces were covered with a high-temperature reusable surface insulation (HRSI) consisting of black tiles. They comprised a silica-based material capable of absorbing the high temperature, acting as heat sinks. The upper surfaces were covered with

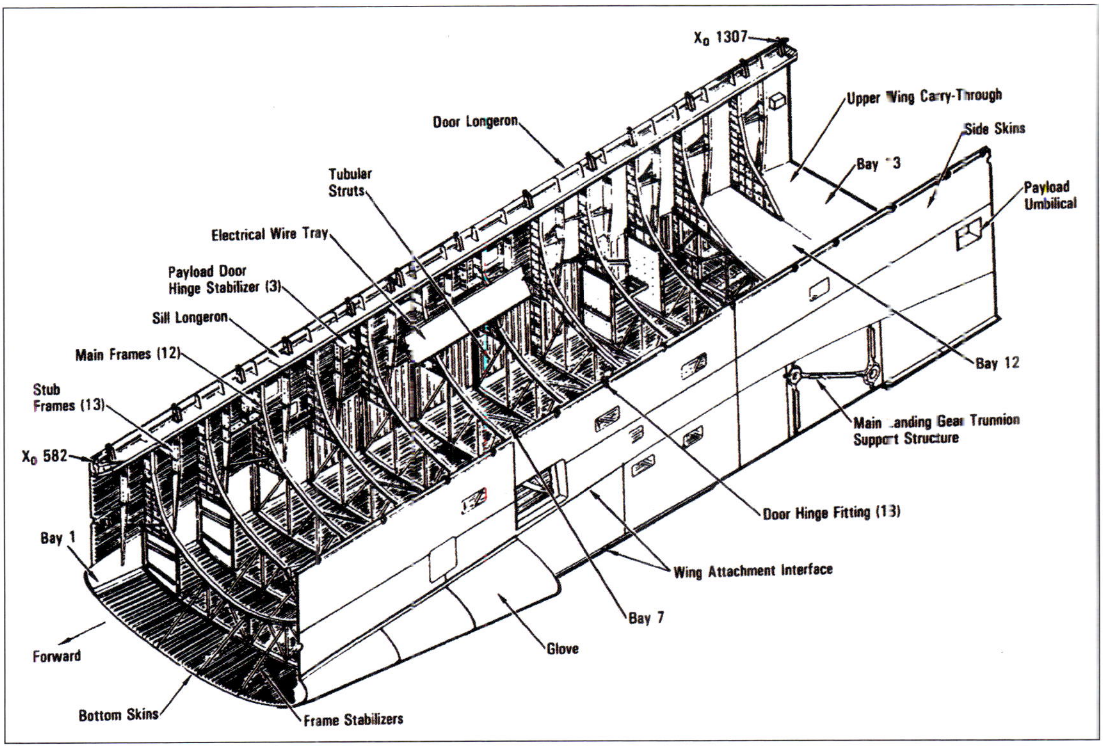

The structural layout of the mid-fuselage section, which transmitted loads from the wings, the aft body and the main rocket engines. (NAR)

low-temperature reusable surface insulation (LRSI) white tiles capable of protecting areas in which temperatures would not exceed 1,260 degrees Celsius (1,200 degrees Fahrenheit).

Overall, there were more than 27,000 tiles, each one uniquely sized for its assigned location, but they were prone to damage and pitting necessitating frequent replacement, a costly and time-consuming job. Over time, most of the white tiles in non-critical areas were replaced with an advanced flexible reusable surface insulation (AFRSI). This consisted of a low-density, fibrous silica batting, a material sewn with a silica thread to create a quilted appearance. Protecting areas where temperatures never exceeded 371 degrees Celsius (700 degrees Fahrenheit), a Nomex felt material was used, saving weight and reducing the number of troublesome tiles.

While external thermal protection solved the problem of recovering the Orbiter intact at the end of each flight, the challenges of maintaining an acceptable temperature in space were no less demanding. At a relatively low average altitude of less than 644km (400 miles), the Orbiter shared equal time between blazing sunlight, with temperatures of more than 93 degrees Celsius (200 degrees Fahrenheit), and deep night, with temperatures as low as -129 degrees Celsius (-200 degrees Fahrenheit). Additional heat came from the internal avionics, electronics and the operation of Orbiter systems, and maintaining a habitable atmosphere.

The Orbiter had an active coolant system with individual radiators on the inner face of each payload bay door, fabricated from a graphite-epoxy composite Nomex material and saving about 23% in weight over an all-aluminum structure. The doors opened with a fold angle of 175.5 degrees, and on the inner face they carried radiators, four panels on each door for a total radiator area of 175m² (1,883ft²). The two forward doors were hinged to fold upward 35 degrees to radiate excess heat from the Orbiter systems and from the Sun's energy when on the daylight side of the Earth.

Freon 21 coolant fluid was passed on demand through 68 tubes on each of the four deployable panels and 26 on each aft panel. Operation of the deployable panels was controlled according to the amount

Above: **Looking aft toward two large boxes of test instrumentation, we can see the interior of the payload bay with the floor yet to be attached. Note the spherical oxygen tanks either side of the payload bay in the immediate foreground. (NASA)**

Left: **The lower section of the mid-fuselage assembly and the payload bay doors, all of which were designed to flex under the extremes of temperature in space. (NAR)**

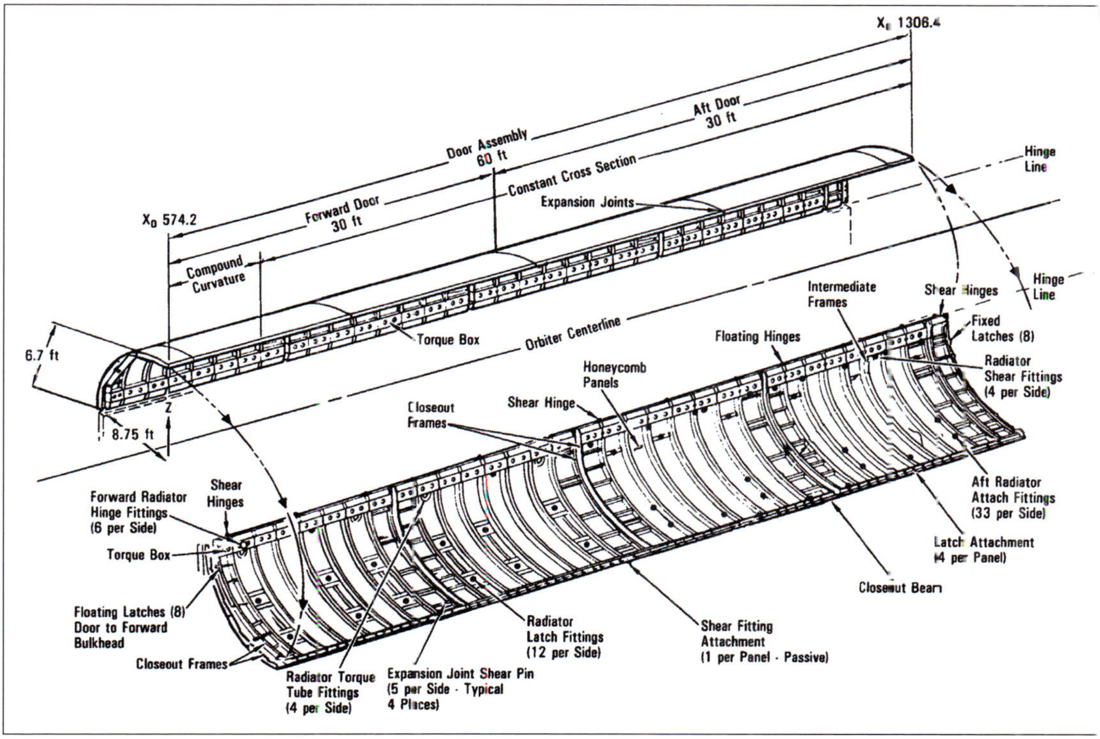

The payload bay doors were secured to the mid-fuselage hinge line and would open shortly after reaching orbit to reveal radiator panels. (NAR)

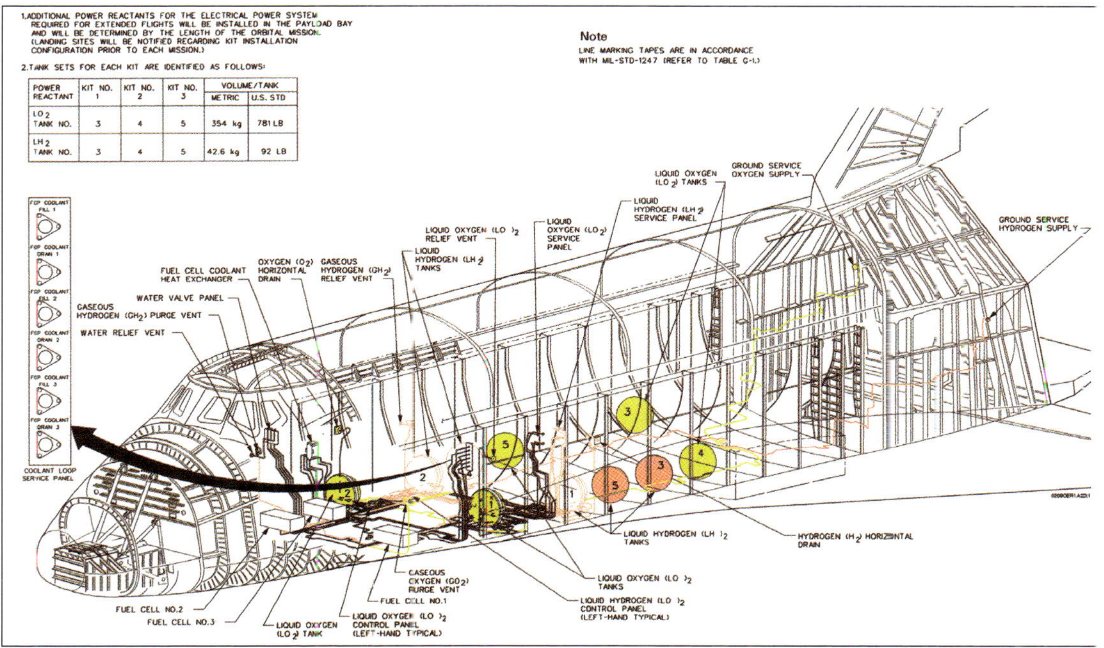

POWER REACTANT	KIT NO. 1	KIT NO. 2	KIT NO. 3	VOLUME/TANK	
				METRIC	U.S. STD
LO2 TANK NO.	3	4	5	354 kg	781 LB
LH2 TANK NO.	3	4	5	42.6 kg	92 LB

This image shows life support and electrical power consumables in the underfloor to the payload bay, together with related systems supported on the floor of the mid-fuselage assembly. (NAR)

of heat required to be radiated to the vacuum of space. The eight radiators could reject up to 30kW of waste heat. It is important to remember that, in space, there is only conduction and radiation, but, in the absence of a noticeable gravity field, there is no convection.

Unlike manned ballistic capsules prior to the Shuttle, which used a pure oxygen environment at one-third sea-level pressure, the crew in the Orbiter had a true "shirt-sleeve" environment. It was a two-gas mix of oxygen and nitrogen roughly in the same proportions and pressure as the atmosphere at the surface of the Earth. There were two reasons for this: the Orbiter would have to support crews aboard the space station, and breathing pure oxygen is not advisable beyond four weeks; and the Shuttle would adopt a "walk-on/walk-off" approach, much like an airline, and had to accommodate non-professional astronauts without special training.

The habitable volume of the Orbiter was a spacious 65.8m³ (2,325ft³) compared with 1.02m³ (36ft³) for the one-man Mercury, 1.56m³ (55ft³) for the two-man Gemini capsule and 6.17m³ (218ft³) for the three-man Apollo spacecraft. Oxygen and nitrogen for pressurizing the Orbiter crew compartment was located in dedicated spherical tanks below the payload bay in the mid-fuselage. It was sufficient for supporting a crew of seven people for 21 days, while giving each crew member almost five times the volume available per person in Apollo.

Crew provision was on a scale completely different to earlier, ballistic manned capsules. Apollo carried provisions for three people across a maximum of 12 days while the Shuttle supported up to seven people for three weeks; 36 man-days versus 147. Food, clothing and personal hygiene items for both men and women were carried in lockers around the walls of the mid-deck area. Initially, sleep provision included bunks, but later missions carried modularized sleep lockers, albeit with little prospect of privacy.

During the Apollo missions, food was provided in tubes, and toilet facilities included nappies and stick-on bags. The Shuttle Orbiter had provision much akin to the flight deck of a cargo aircraft, with facilities below for rest, eating and a primitive toilet known as the waste management system, designed for both sexes. From June 1983, women astronauts were frequent assignees on Shuttle missions, as crew members and pilots but also as commanders in the left seat.

Power to make things move included a hydraulic system, auxiliary power units (APU), and rocket propulsion. Located in the aft fuselage, three APUs powered the hydraulic system in an arrangement

***Above left*: Essential for directional control during independent atmospheric flight, the vertical fin was built by Fairchild and consisted of a support for the split-rudder assembly. (NAR)**

***Above right*: Grumman built the wing for the Orbiter, with a thickness/chord ratio of about 8.3% at the glove root. (NAR)**

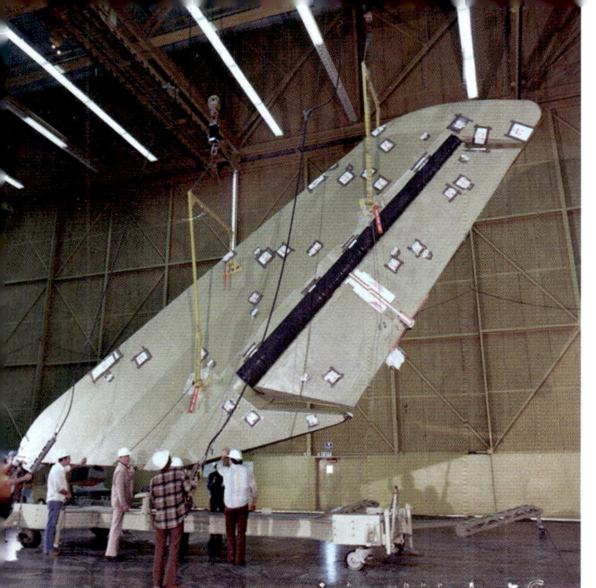

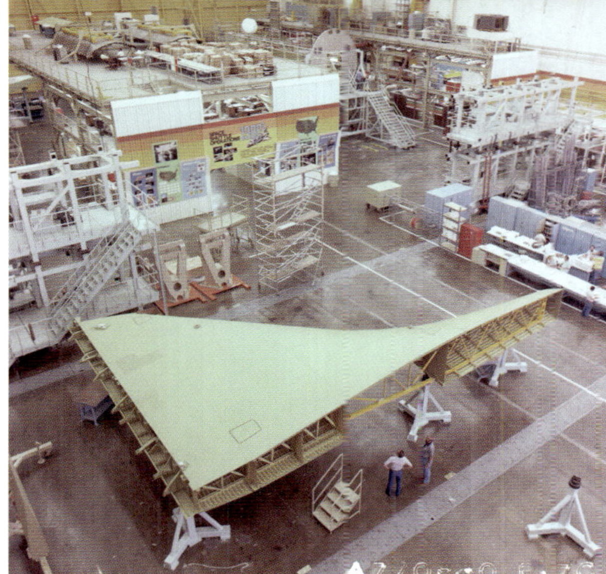

similar to that on large commercial aircraft. They achieved power through the decomposition of hydrazine over a catalyst to produce hot gas and were operated from five minutes before launch until the Orbiter reached orbit, when they were shut down. One APU would be activated daily to verify operability, and all three restarted after the de-orbit burn to control the aero-surfaces and deployment of the landing gear.

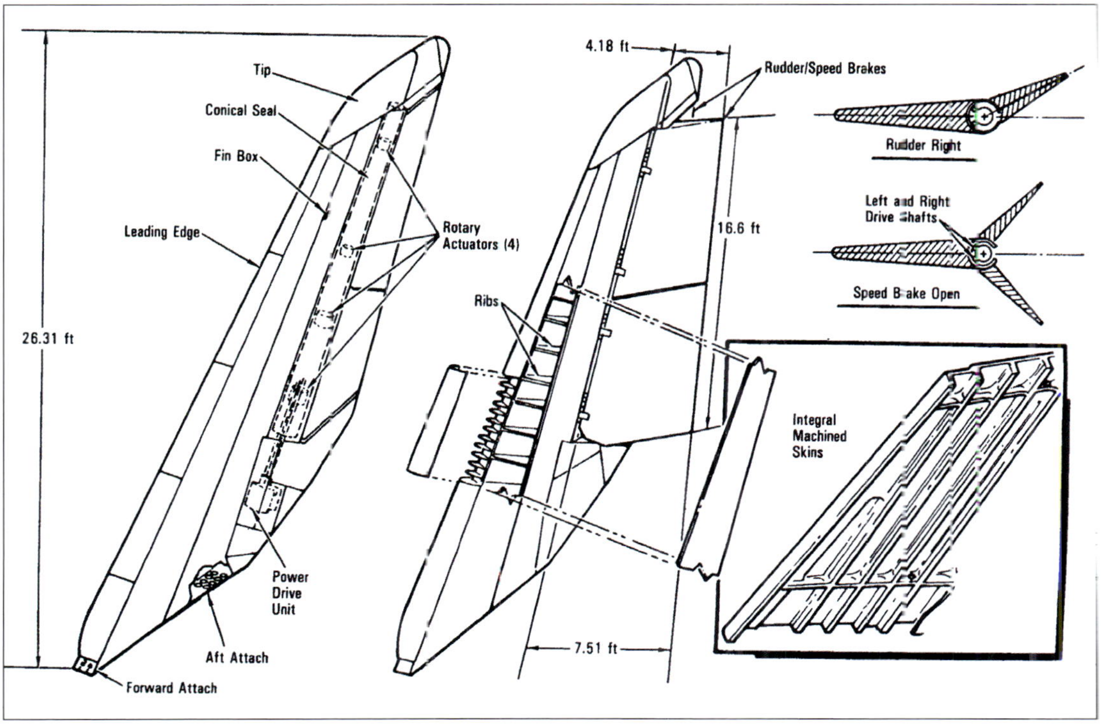

The Orbiter rudder would double as a speed brake. (NAR)

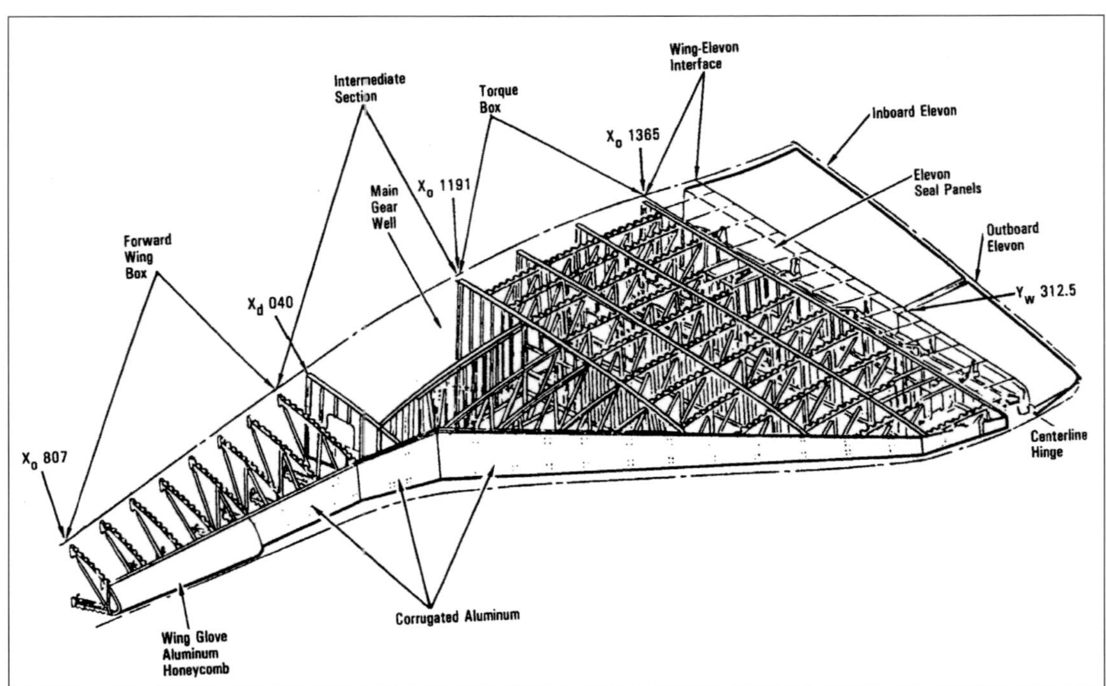

The wing provided a broad torque box with aluminum facing panels for thermal protection on the leading edge, and support for flight control surfaces along the trailing edge. (NAR)

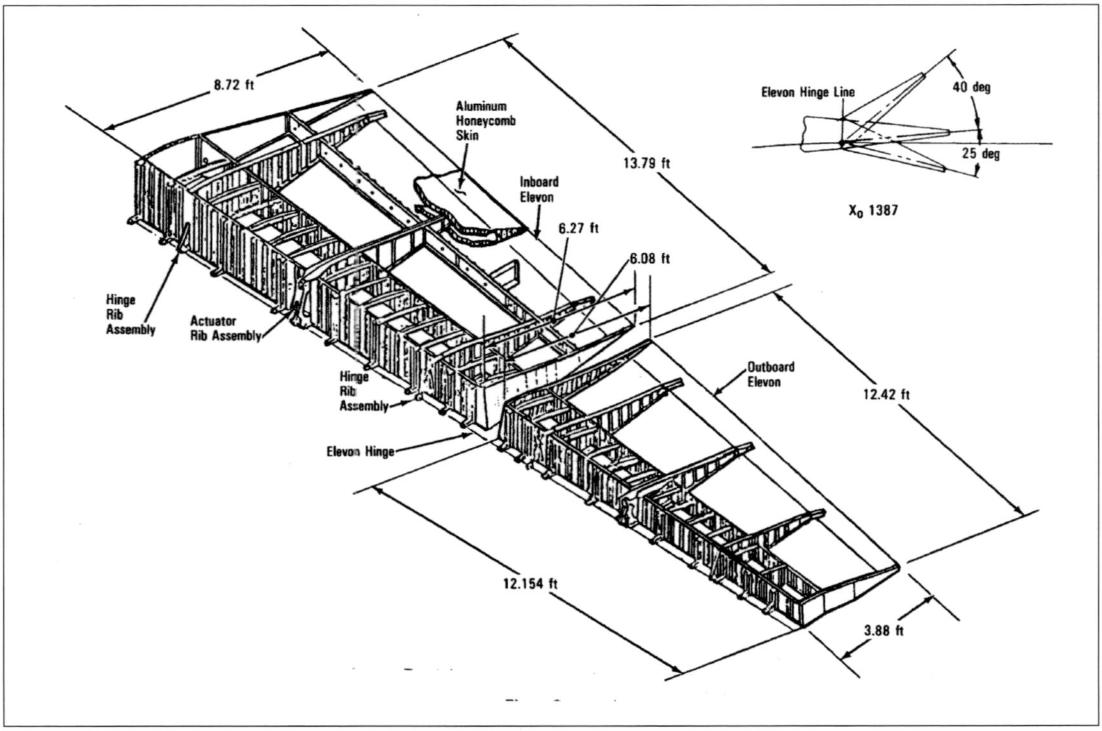

Inboard and outboard elevons controlled roll and pitch and occupied almost the entire wing trailing edge. (NAR)

The hydraulic fluid was a synthetic hydrocarbon to a military specification. It controlled the Orbiter's aerodynamic control surfaces, the body flap, landing gear doors, main landing gear deployment (no retraction being required), and the closeout door to the propellant feed conduits on the underside of the fuselage. The system also operated several functions on the three main rocket motors, controlling various valves on the pre-burners and main engines. Hydraulic power was also used to gimbal the engines and so maintain the required thrust alignment with the planned trajectory after the solid rocket boosters (SRBs) were jettisoned.

Electrical power relied on three fuel cells, a technological first applied to the two-man Gemini spacecraft and flown since August 1965. It was the only means of generating electrical power on the Apollo spacecraft. Operating by reverse electrolysis, bringing hydrogen and oxygen together over a potassium hydroxide catalyst to produce electricity, the fuel cell has proven to be both functional and extremely reliable. But it comes at the cost of complexity compared with alternate methods of power production. Because the Shuttle required a continuous power supply of around 21kW at 28 volts dc, batteries alone would be too heavy, and solar arrays would only work on the daylight side of Earth.

Called reactants, cryogenic hydrogen and oxygen were stored in spherical containers below the payload bay, in liquid form to shrink their volume. Oxygen is stored at -182 degrees Celsius (-297 degrees Fahrenheit) and hydrogen at -254 degrees Celsius (-425 degrees Fahrenheit). More than twice as much liquid hydrogen than gaseous hydrogen can be contained in a given volume; liquid oxygen is 1,000 times denser than gaseous oxygen. Moreover, the principle also produces water as a by-product, and that is used to cool the fuel cells and to flow into the Orbiter cooling system as well as being available for

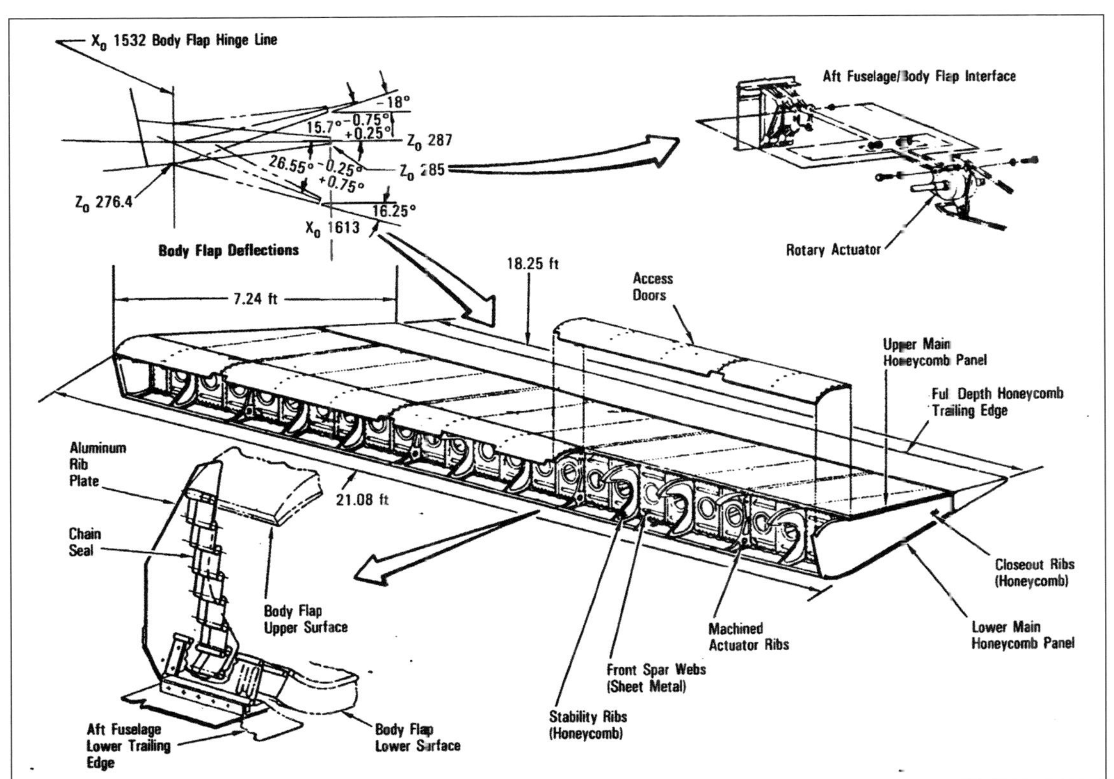

Attached to the aft fuselage, the body flap was originally incorporated to protect the main engines from heat during re-entry but was utilized for pitch trim during descent and landing. (NAR)

Above left: Here we see the main landing gear leg, with retraction conducted by ground crew prior to mating with the ET. (NASA)

Above right: The nose wheel well and upper leg attachment points. (NASA)

rehydrating food and for purification for drinking. The number of tanks installed depended upon the duration of the mission, but the same oxygen supply was also used to pressurize the crew compartment.

Rocket motors and boosters

The key advantage for the Shuttle was that it inherited the substantial experience with cryogenic propellants for highly efficient rocket motors. These offered more payload lifting capacity per unit mass of propellant than for any other chemical combination in use, far exceeding the efficiency of oxygen and kerosene or storable combinations such as hydrazine and nitrogen tetroxide. Put crudely, they delivered more "bang for the buck!" It was for this reason that cryogenic second and third stages were selected for the giant Saturn V Moon rocket and for the Centaur upper stages used to launch satellites, unmanned lunar robots, and planetary probes.

The US had a lead in cryogenic rocket motors and in handling large quantities of hydrogen and oxygen propellants from the mid-1950s. It was triggered when British engineer Raymond Rae convinced US manufacturers to plan for large-scale cryogenic production and storage for what he envisaged to be the dawn of hydrogen-fueled aircraft engines. The government backed research into large-scale production, but when the aircraft engine idea evaporated, cryogenic propellants for rocket motors became not only suitable but preferable. So it was that the cryogenic industry was in place when rocket engineers turned to this most efficient of propellant combinations to squeeze the maximum performance from the minimum weight.

Nevertheless, demands on the Space Shuttle Main Engine (SSME) were great. Very few reusable rocket motors had been designed and operated. Perhaps most notable being the XLR99 from Reaction Motors Incorporated for the North American X-15 research aircraft between 1959 and 1968, although that was not a cryogenic motor. Moreover, although it was throttleable, it had a maximum thrust of only 253.5kN (57,000lb). At around 2,000kN (450,000lb), the optimum thrust level of the SSME would be more than six times greater. The XLR99 engine was throttleable between 50% and 100% but unstable at lower settings. The SSME could be throttled in the range 67–109% across a firing time of more than eight minutes, compared with around 1min 20secs for the XLR99.

In reality, there was no comparison between these two engine programs. The SSME being infinitely more complex, efficient and with a more demanding specification. A more effective comparison can be made with the Rocketdyne J-2 engine used in the second and third stages of the Saturn V. With a fixed thrust of 1,112kN (250,000lb) and variable only through tweaking the mixture ratio during flight. The cryogenic J-2 had a maximum operating time of approximately six minutes when operating in a cluster of five motors on the second stage, and two burns totaling about 8mins 30secs operating as a single engine on the third stage.

The technical challenges and the design requirements raised the bar on engineering solutions. For instance, to achieve optimum performance, the SSME combustion chamber was built to operate at a pressure of 8.95kg/cm^2 (3,000lb/in^2) compared to 2kg/cm^2 (700lb/in^2) for the J-2 in the Saturn program. Very few rocket motors operated at chamber pressures greater than 2.98kg/cm^2 (1,000lb/in^2). In addition, the SSME leaned heavily on the pre-burner concept where propellants are burned prior to entering the combustion chamber to gain an advantage from configuring the chemistry of the propellants. This also enabled some hydrogen and oxygen to drive the turbine pumps for delivering the propellants to the combustion cycle.

The contract to build the SSME was won through a competitive bid between Rocketdyne and Pratt & Whitney, a process that began long before the formal decision to build the Shuttle. Over time, the requirement changed significantly. During initial design studies, NASA expected to build a two-stage system with the manned fly-back booster powered by 12 rocket motors and an Orbiter powered by two

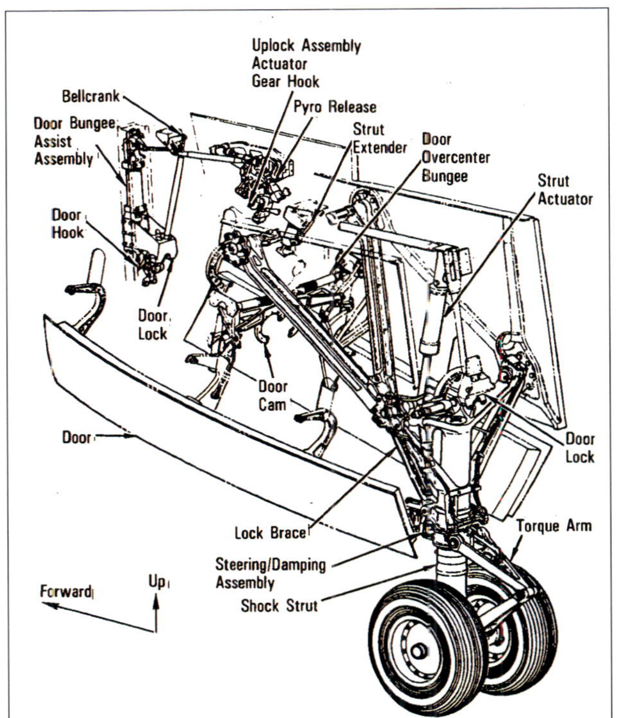

Above left: **Gear deployment was usually achieved by hydraulically driven actuators, but, if that failed, it could be completed by gravity alone. (NAR)**

Above right: **The leg was designed for simplicity, but some problems were encountered with the tires, and changes were necessary during the program. (NASA)**

or three. As payload requirements changed and the decision was made to downsize the grand vision of a two-stage manned vehicle with its complex fly-back booster, the specification for the SSME was refined.

Rocketdyne won the contract in 1971, a year before the definitive design configuration of the Shuttle had been finalized. By this time, there would be no manned fly-back booster and the Orbiter would have three, not two main engines. The need for reusability pushed the technology, but this was a crucial element in justifying the Shuttle itself. Engines were expensive. Recovering them intact and returning them to flight with only modest rework was a key element in the economic value proclaimed for the entire concept.

Where once the propellants were to have been stored in tanks within the Orbiter, now they would be contained within the External Tank (ET), a cylindrical structure 46.8m (153.8ft) in length with a diameter of 8.4m (27.6ft). The liquid oxygen would be stored above the liquid hydrogen in tanks fabricated, at first, from 2219 aluminum. Later, tanks would be fabricated in 2915 aluminum-lithium, which was lighter and stronger. Propellant was delivered to the Orbiter's SSMEs through two separate feed lines, each 43cm (17in) in diameter, via interconnects on the base of the Orbiter aft fuselage with closeout doors shut when the Orbiter reached space.

A special CPR-488 spray-on foam insulation (SOFI) would help inhibit soak-through from the hot Florida sun and insulate the cryogenic tanks while on the pad and during powered ascent. Extensive weight-saving methods were applied continuously throughout the operational life of the program, as flight data showed where improvements could be made to reduce mass and enhance payload capabilities. The ET was the only element of the Shuttle that was expendable, although some studies were conducted on its potential use for building space station modules with tanks carried all the way into orbit.

In its primary function, the ET was more than just an assembly of propellant tanks. It formed the backbone to which the twin SRBs and the Orbiter were attached. It accepted and transmitted loads from the SSMEs in the Orbiter as they ignited on the pad several seconds before ignition of the SRBs, an event that signaled lift-off, and then transmitted loads from the severe vibration of the boosters for just over two minutes until they were jettisoned. After that, the ET would continue to provide propellant for a further seven minutes.

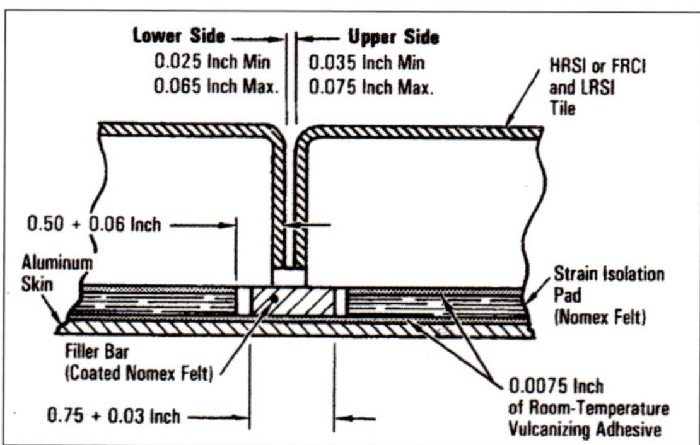

Above left: Thermal protection for the Orbiter during re-entry consisted of black and white reusable tiles and reinforced carbon-carbon (RCC) for places likely to experience the highest temperatures. The RCC nose cap contrasts with the black under-surface tiles. (NASA)

Above right: Brittle tiles would craze and crack under thermal expansion with cycling temperature extremes, so more than 30,000 tiles were tailored for each particular location, with expansion joints to accommodate minor flexing of the aluminum structure to which they were attached. (NAR)

At first, the ET had a small rocket motor in the nose which would deorbit the inert tank and bring it back down over the Indian Ocean, but that changed. As finally configured, the ET would provide power to the Orbiter just short of orbital velocity, separate, and leave the final push to be provided by two short burns of the Orbital Maneuvering System (OMS) engines situated in pods either side of the vertical tail. The OMS engines had their own supply of monomethyl hydrazine and nitrogen tetroxide propellants with gaseous helium used to pressurize the tanks and force the fuel and oxidizer into the motors on demand. The engines were fixed in the installed position, and each had a thrust of 26.67kN (6,000lb) with the capacity to provide the final push into orbit, to carry out major orbital changes and to provide a deorbit burn to start the Orbiter down toward the atmosphere.

Each OMS pod also provided support for 14 attitude control thrusters, which would have their own dedicated propellant supply separate from the OMS engines so that any failure of either system would not inhibit the other. The 12 primary thrusters would each have a thrust of 3.8kN (870lb), with two vernier thrusters producing a thrust of 106N (24lb) for fine tuning attitude excursions such as docking with the space station modules. Each thruster could be operated between 1sec and 125secs and were designed to operate over 330,000 cycles before replacement. The attitude control system had 14 additional thrusters of an identical type in the forward fuselage, of which four were verniers.

Serviceability, ease of access to systems and subsytems, durability, and long life were fundamental aspects of the Shuttle configuration, and that extended to servicing and general maintenance duties between flights. The OMS pods, for instance, were bolted to the aft fuselage as integral units so that after each flight they could be quickly disconnected, taken aside and worked on, leaving other technicians to work with the aft sections of the Orbiter without interference. One of the big lessons learned from the Shuttle program, however, was the extended time required between flights to turn each vehicle around and get it back on the pad for the next mission.

The decision to opt for solid propellant rocket motors to boost the Shuttle off the pad was a significant shift in philosophy. It had always been said that NASA would never put astronauts on solid propellant rockets. Once started, they could not be shut down, and their thrust could not be controlled, staying instead at a level that would be fixed by the chemistry of the solid propellant and by the way it was contained within its cylindrical case. Solids are much less efficient per unit weight of propellant than even the basic liquid propellants such as oxygen and kerosene, much less so than the cryogenic propellants carried in the ET.

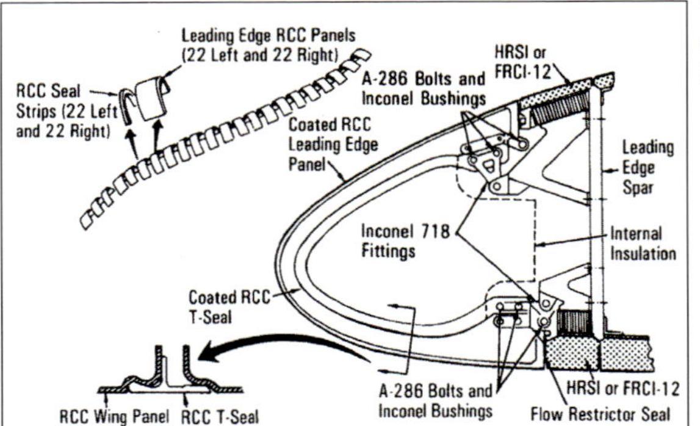

The RCC panels shaped the contoured leading edge of the assembled wing, the panels being bolted to the leading edge spar. (NAR)

Left: An RCC panel is being installed on the leading edge of the glove, one of 22 on each wing to protect against temperatures of up to 1,648 degrees Celsius (3,000 degrees Fahrenheit). (NASA)

Below: A lot of places around the Orbiter could produce heat from the avionics and various operating systems, so thermal energy was removed by a system of coolant loops and carried to radiators on the payload bay doors. (NAR)

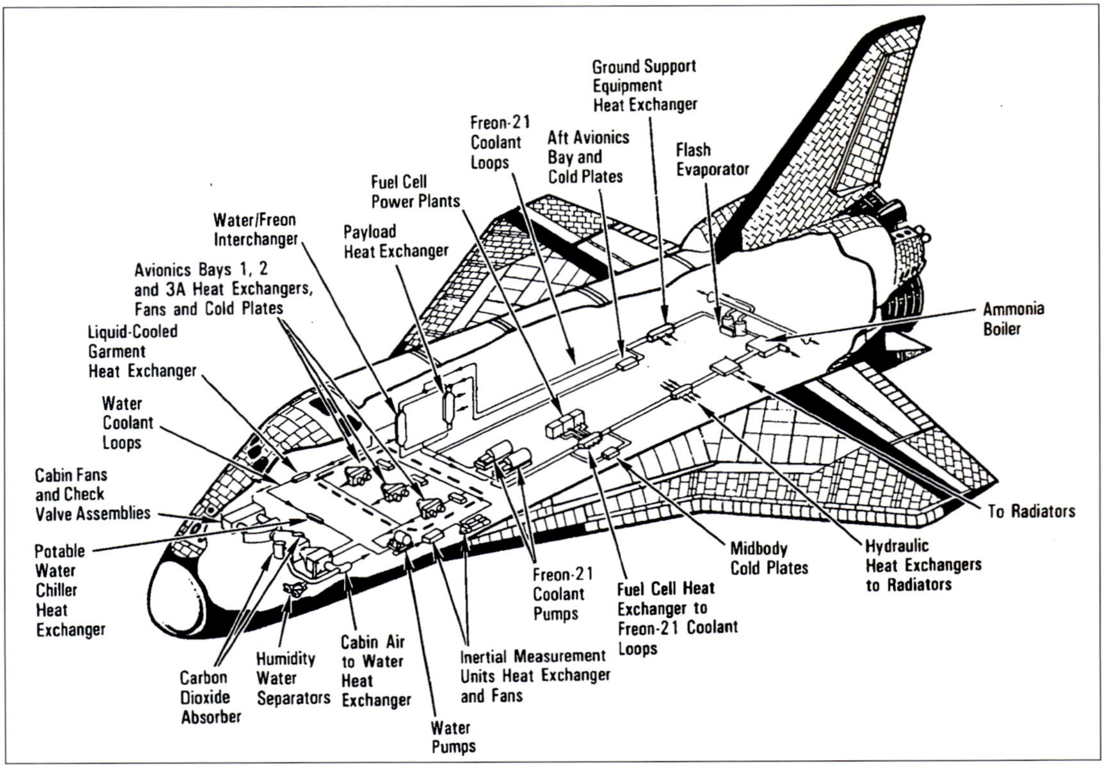

However, the increased confidence in solids came from their widespread use for *Minuteman* and *Polaris* ballistic missiles and as boosters for Titan rocket satellite launchers. In these programs, they were proving statistically just as reliable as liquid propellant engines, which can be shut down and controlled on demand. Nevertheless, they would break new ground in thrust for production rockets of this kind.

The mighty *Minuteman* ICBM had a thrust of 792kN (178,000lb), but the two solid rocket motors for the Titan IIIC satellite launcher each had a thrust of 5,204kN (1.17 million lb) with later versions introduced in the mid-1970s delivering 6,227kN (1.4 million lb). For early space missions, each Shuttle booster would be required to deliver a thrust of 12,454kN (2.8 million lb) at launch, and later versions

Right: A fixed radiator section was attached to the aft payload bay doors with deployable radiators forward, their operation depending on the requirements of a particular orbit. (NAR)

Below: The environmental control requirements of the Orbiter were modeled on the military and commercial large aircraft fleets with a range of optional fitments and fixtures in both the flight deck and the mid-deck areas. (NAR)

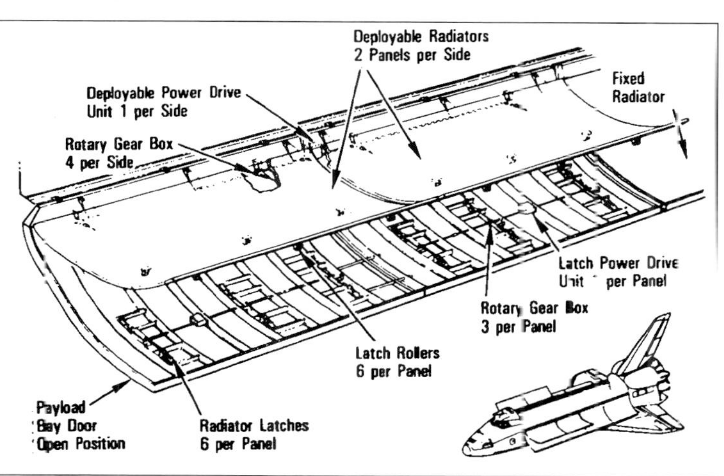

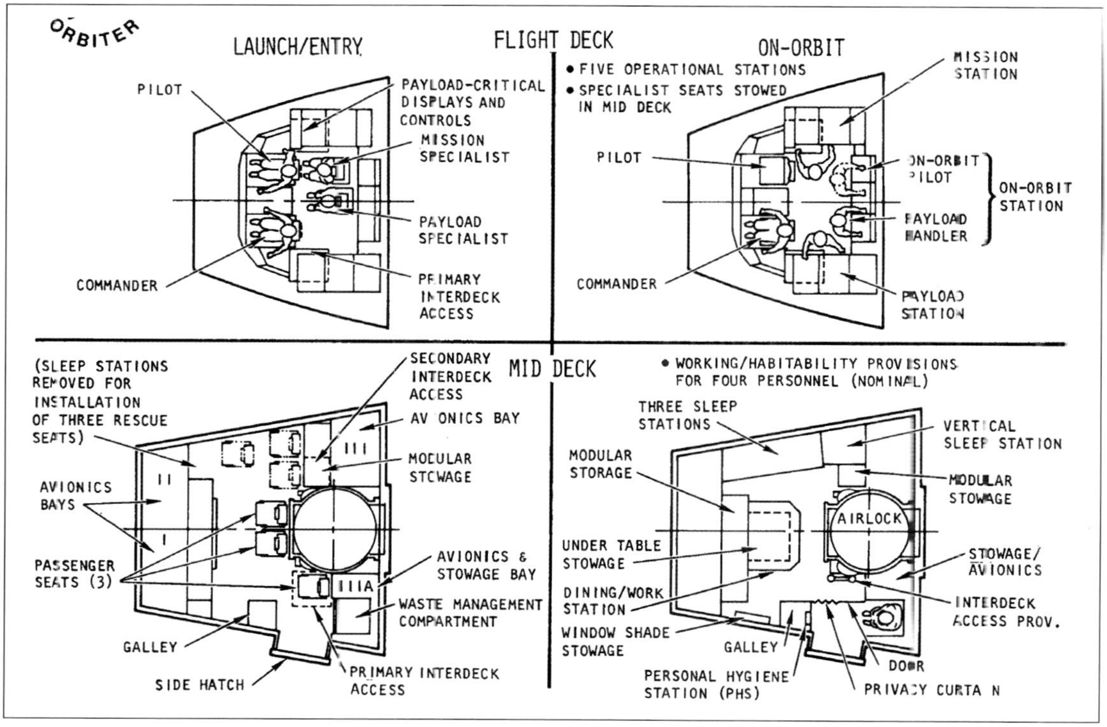

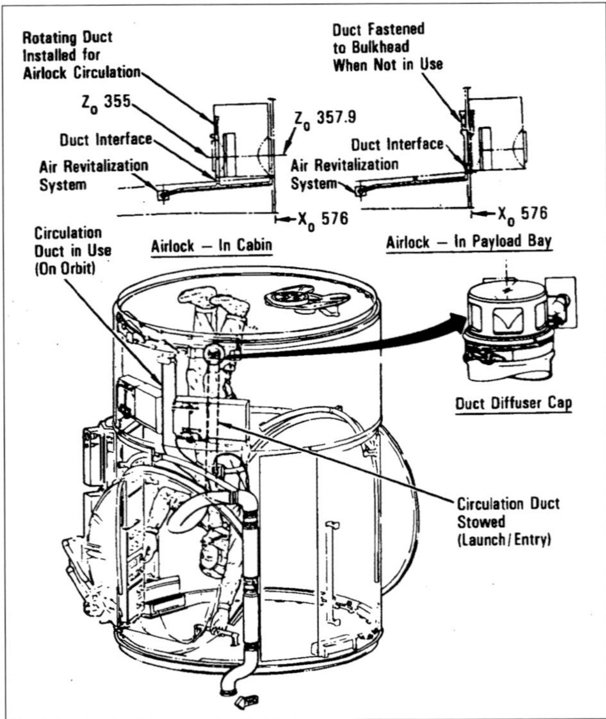

would produce 13,344kN (3 million lb) and run for about two minutes before burnout and separation. In all cases, thrust would diminish with time to compensate for the loss of weight as propellant was consumed in the SRBs and the SSMEs. This ensured acceleration was kept below 3g.

The economics of procurement cost, production and operation and their simplicity of operation tipped the scales in favor of solids. In November 1973 – nearly

Left: Initially, the airlock was in the mid-deck area, but on later missions it was outside in the payload bay. It placed additional demand on environmental control requirements, and it was an essential access point for space-walking astronauts working inside the payload bay. (NAR)

Below: The three hydrazine-fueled Auxiliary Power Units (APUs) operating the hydraulic systems and major elements (including the water boiler) were located in the aft fuselage assembly. (NAR)

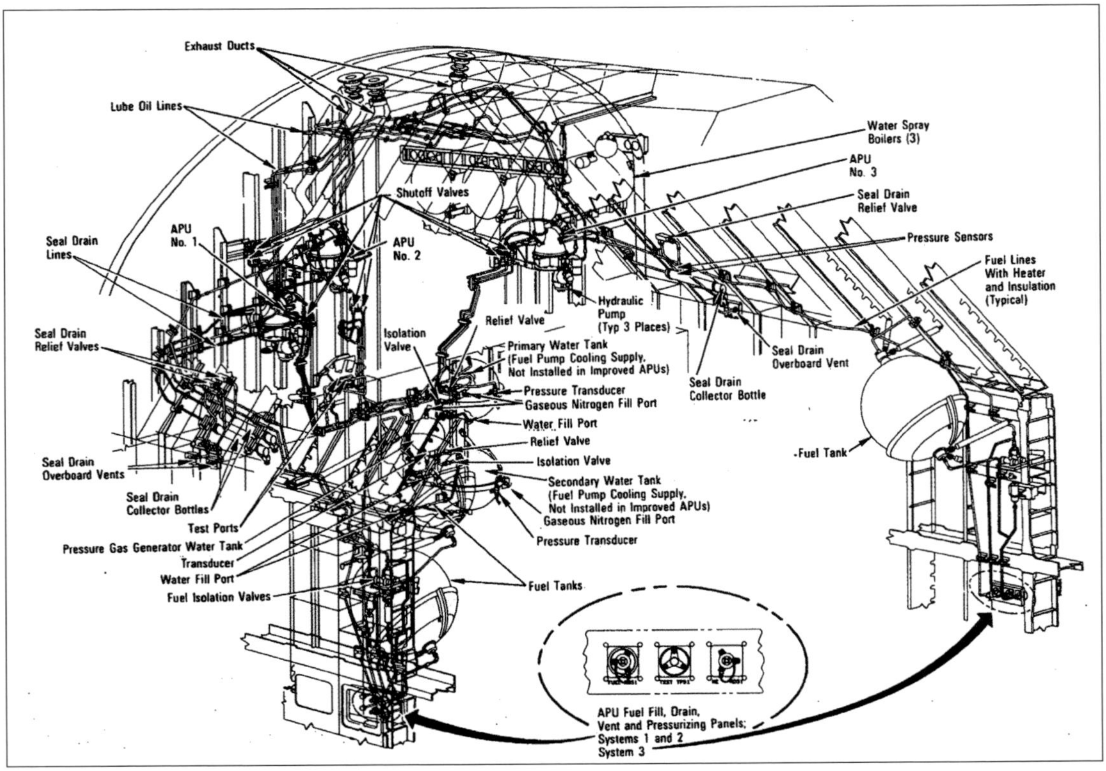

two years after the Shuttle had received presidential approval – NASA placed a contract with Thiokol for the production of the most powerful solid rocket motor (SRM) ever built for operational use. A month later, United Technologies was given the job of producing the solid rocket booster (SRB), the cases, electronic control systems and separation thrusters designed to push them away from the ET. It was the last of the major contracts for the Shuttle.

There were two approved methods of producing a large solid rocket booster: a single case filled with propellant, known as the monolithic concept; or a series of barrel-shaped segments bolted together to form a long cylinder. Each would require an exhaust nozzle and control system, and in the case of the Shuttle solids, parachute recovery systems so that they could be retrieved after splashdown and refilled for further use. Both types would have to incorporate recovery systems.

NASA chose the segmented concept for two reasons. With a projected length of 45.45m (149ft) and a diameter of 3.71m (12.17ft), a monolithic cylinder could not be moved by road or rail because the radius of curvature approved as standard in the US was too tight. And with reusability at the forefront of decisions, it would be less costly to produce a new segment than a completely new booster if part of the SRB was found to be unusable after splashdown. But the segmented concept had its own safety concerns: when ignited, the internal shock wave would cause a booster of such enormous thrust to momentarily expand out like a balloon, imperceptible to the human eye but very real, before returning to its original shape.

To accommodate that, there would have to be a highly effective seal between the segments to prevent a flash of combusted gases escaping through the gap between them and working like a blowtorch on the supports for the ET or the Orbiter. It was precisely that which brought about the destruction of *Challenger* on 28 January 1986, when cold temperatures made the rubber O-rings less flexible in the tang and clevis joint connecting adjacent segments. It was precisely this risk that NASA accepted when agreeing to the segmented concept.

Each SRB consisted of a nose section, the SRM and the aft assembly where most of the electronics, the hydraulic system and the nozzle with its thrust-vector controls were located. The SRM comprised 11 D6AC steel sections stacked one on top of the other into four separate castings, or motor segments, each joined to the next by 177 pins spaced equally around the cylindrical section. With an empty weight of 91,000kg (200,000lb), each SRB held 499kg (1.1 million lb) of propellant, a composite of 15% aluminum powder for fuel, 69.6% ammonium perchlorate as the oxidizer, 0.4% iron oxidizer powder as a burn rate catalyst, 12% of a rubber-based PBAN binding agent, and 2% of an epoxy curing agent.

The propellant was poured into each segment in a semi-liquid state around a star-shaped mandrel placed down the center. When the propellant had dried to a solid, the mandrel was withdrawn, leaving the mirror impression of a star-shaped cavity down the center. Electrical power

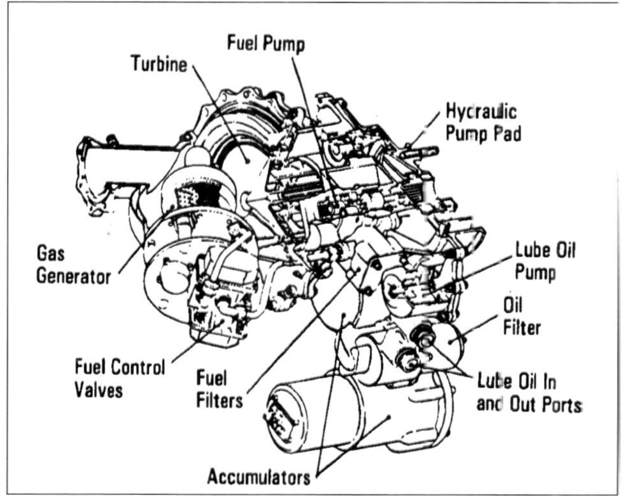

Three APUs provided power for the three standard 20,685kPa (3,000lb/in²) hydraulic systems on a redundancy basis, ensuring operation for engine, aero-controls and landing gear deployment actuators. (NAR)

to operate the SRBs came from the Orbiter fuel cell system with hydraulic power units powered by hydrazine, much like the APUs in the Orbiter, for controlling actuators that move the nozzle for directional control during the powered phase. The SRBs also carried a self-destruct system, which enabled ground controllers to activate linear shaped charges and break open the structure to prevent them careening into populated areas. This was activated only once in the program. It occurred when the Shuttle broke apart shortly after launch on January 28, 1986.

With both boosters still on full thrust, there was a danger the uncontrolled SRBs would fly back across the Florida coastline and head inland.

Preparations for flight began when the booster segments together with the fore and aft sections were delivered to KSC by rail and stacked in the Vehicle Assembly Building (VAB). On one standard delivery assignment, rail cars jumped the tracks and some segments tipped off and fell 3m (10ft), but very few incidents like that occurred, and the routine supply of SRB elements went forward as planned.

When stacked together vertically in the VAB at KSC, the separate sections line up to present a continuous cavity from top to bottom. Assembly of the entire Shuttle begins with the segments stacked and bolted together on the Mobile Launch Platform (MLP) in the VAB, to which the ET is inserted between them, followed by the Orbiter to the ET. When ready for launch, the stack was rolled out to the pad by the Crawler Transporter (CT). The VAB, the MLP and the CT were developed for the Apollo program and date back to the early 1960s in design, as did the pads – Launch Complex 39A and 39B.

At the assigned pad, the assembly was given a final series of checkouts and tests with all elements. Ignition took place when the Orbiter engines reached 90% thrust. A solid propellant igniter, 1.14m (45in) in length, sent a flash down the entire center of the booster beginning combustion from the inside out. The star-shaped cavity controlled the burning process and was designed to reduce thrust by around 35% at 55secs. To prevent the outer case burning through, an inhibitor lined the inner face.

When the SRBs burned out at an altitude of about 46km (28 miles), explosive bolts severed the connections to the ET, and four solid rocket thrusters in the nose and in the base of each SRB fired for one second to push the inert cases aside as the ET and the Orbiter continued on up into orbit. The boosters would continue to ascend to a maximum altitude of 67km (41.6 miles), from where they then began to fall back through the denser regions of the atmosphere.

At an altitude of about 4,785m (15,700ft), two pilot parachutes would deploy, which in turn extracted two 16.45m (54ft) drogue parachutes followed by three 42.45m (136ft)-diameter main parachutes, a process slowing the booster from 142kph (230mph) to 80.45kph (50mph). Entering the water in a

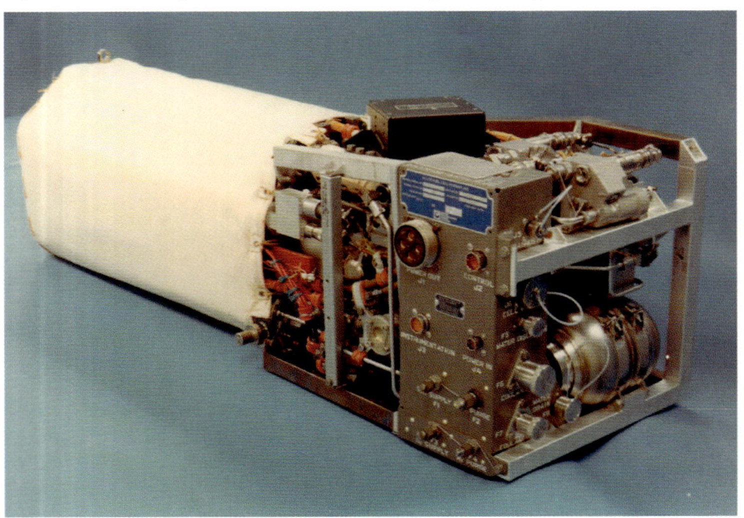

vertical orientation about 4mins 39secs after lift-off, divers would drop from a helicopter and insert a plug in the nozzle, pump out air and residual water, which would turn them 90 degrees to lie flat on the surface for towing back to Port Canaveral.

The three oxygen/hydrogen fuel cells provided 28-volt dc systems for the Orbiter and secondary systems attached to the three-bus system. (NASA)

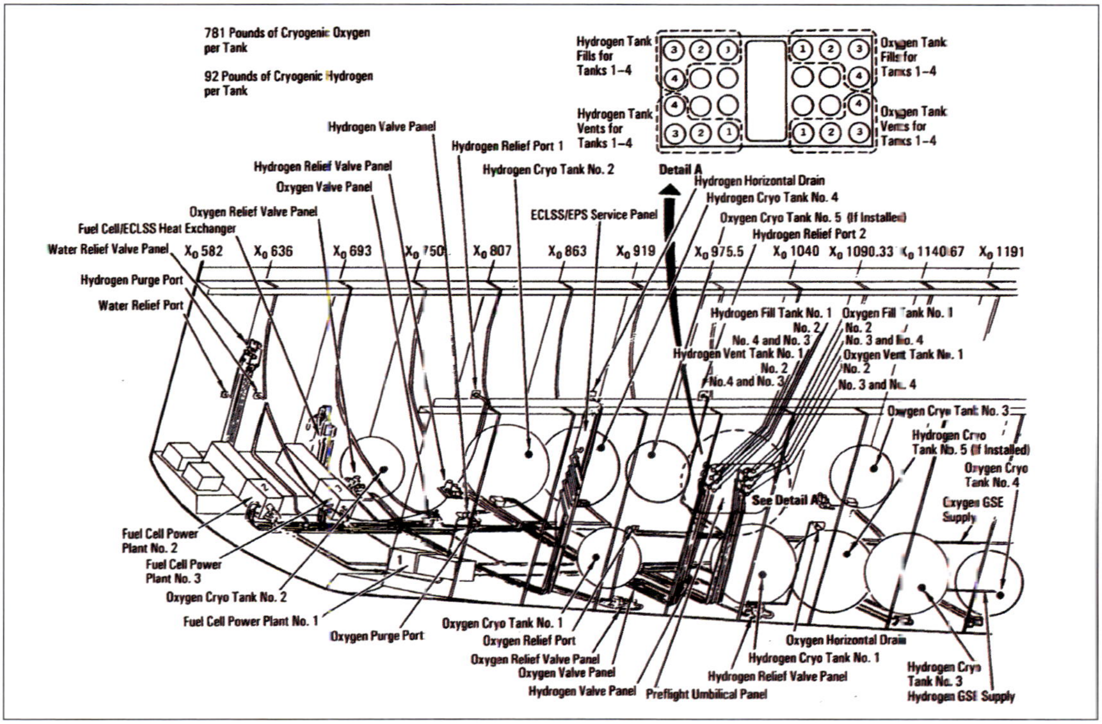

Hydrogen and oxygen reactants were stored in two-tank sets located beneath the floor of the payload bay, with varying numbers according to the requirements of a mission, a normal seven-day flight calling for four sets. Five could be accommodated if necessary. (NAR)

Above left: NASA ran 14 extended duration missions for which additional reactant sets of paired hydrogen and oxygen tanks were stacked, as shown here in preparations for the first, the 14-day flight of STS-50. (NASA)

Above right: When the additional eight reactant tanks had been attached to the back of *Columbia*'s payload bay for STS-50 in June 1992, a thermally protective case was placed over it. The longest mission was STS-80 in late 1996, lasting 17 days 15 hours. (NASA)

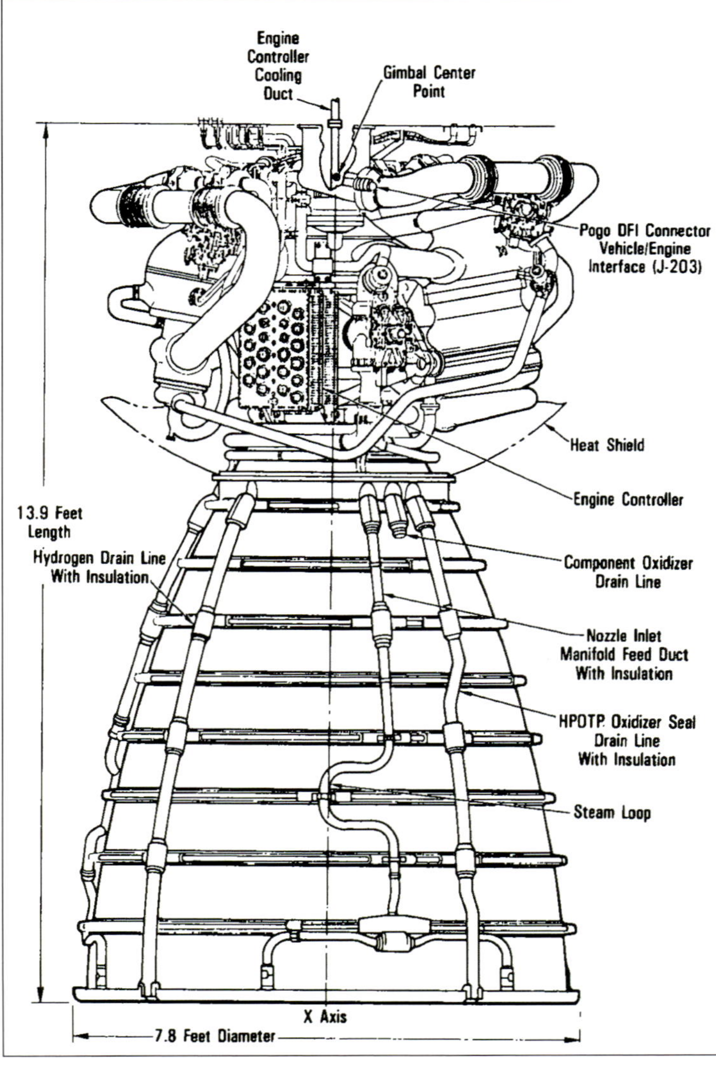

Above left: The Space Shuttle Main Engine (SSME) powered the Orbiter all the way from the launch pad to near-orbital velocity, but many improvements were made as the program evolved, particularly to the turbopumps. (NASA)

Above right: The aft fuselage supported the three SSMEs, the two Orbital Maneuvering System (OMS) engines, and a set of attitude control and translation thrusters. (NASA)

Left: The SSME was built around a high-pressure combustion chamber with preburners for the cryogenic propellants, throttleable, and with self-diagnostic engine management electronics. (Rocketdyne)

Key to the efficiency of the high-pressure combustion chamber were the preburners, which also drove the turbopumps. (Rocketdyne)

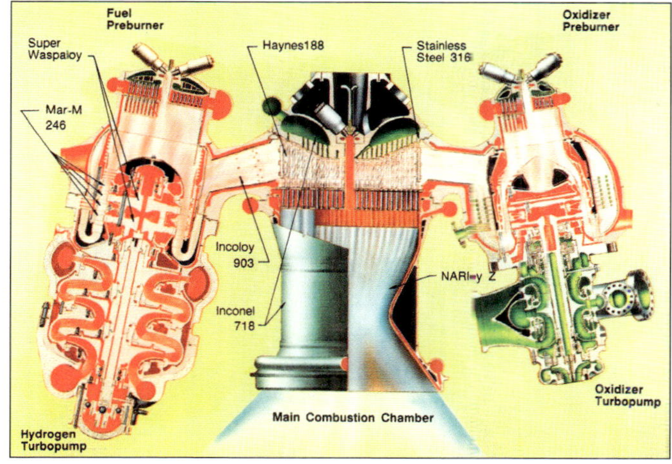

Right: The first ET was rolled out by Martin Marietta on June 29, 1979, and would be used for the first Shuttle launch on April 12, 1981. (Martin Marietta)

Below: The ET comprised two separate tanks for the liquid oxygen and liquid hydrogen and delivery lines to the Orbiter engines. (Martin Marietta)

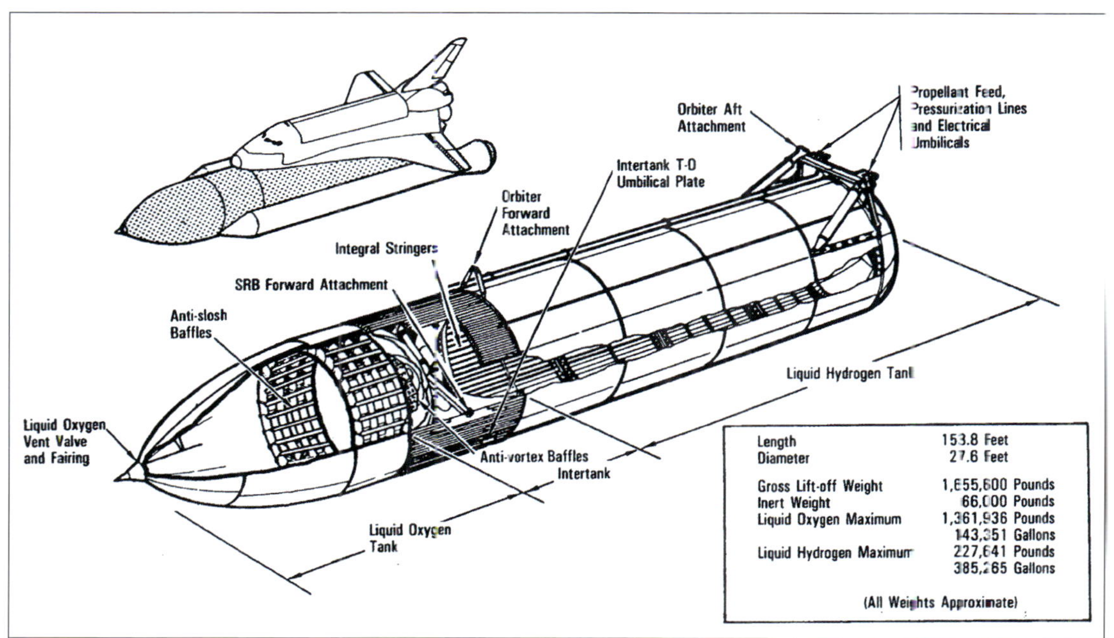

Length	153.8 Feet
Diameter	27.6 Feet
Gross Lift-off Weight	1,655,600 Pounds
Inert Weight	66,000 Pounds
Liquid Oxygen Maximum	1,361,936 Pounds
	143,351 Gallons
Liquid Hydrogen Maximum	227,641 Pounds
	385,265 Gallons
	(All Weights Approximate)

Above left: This image shows the ET with its spray-on foam insulation applied, the aft attachments for the Orbiter, and the propellant transfer fixture. The long LOX transfer pipe can also be seen extending down the length of the tank as well as the smaller systems tunnel. (NASA)

Above right: A technician applies the spray-on foam insulation (SOFI), which will further insulate the cryogenic tank. (NASA)

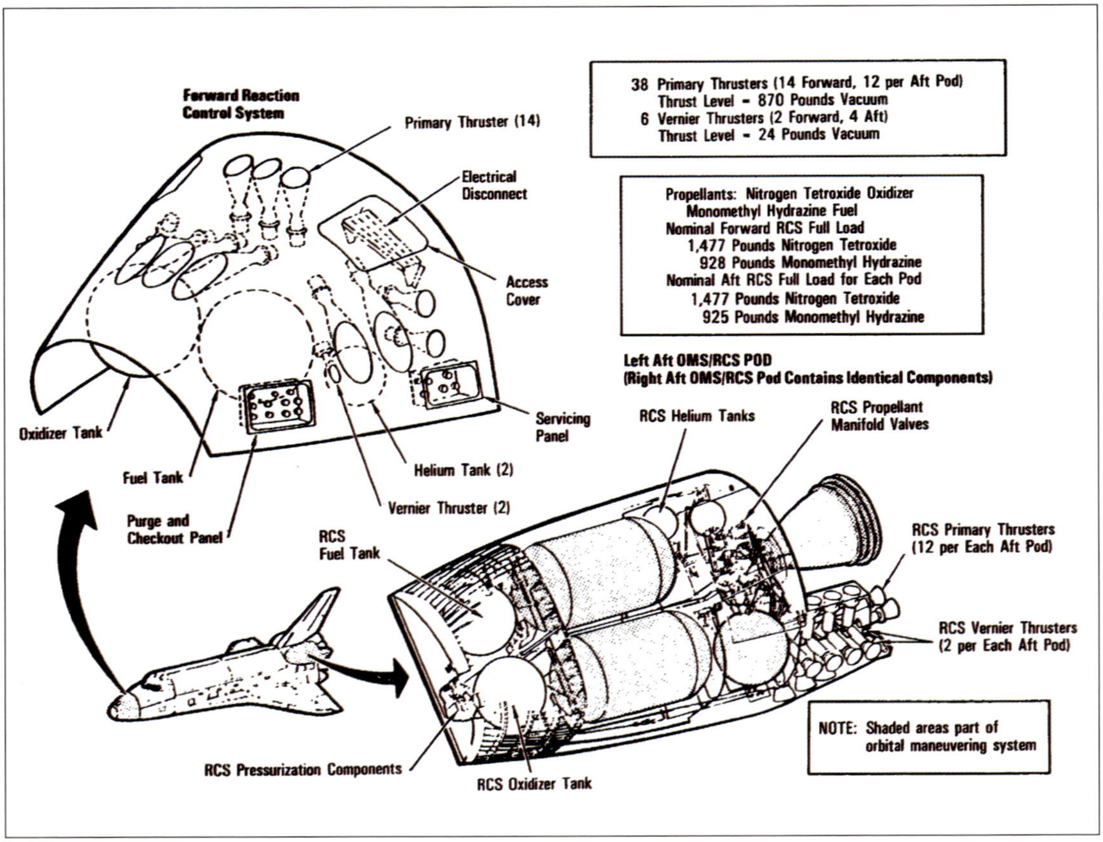

The aft propulsion pods carried the OMS engines, propellant and the aft attitude control, and translation thrusters. (NAR)

Above left: Displaying propellant tanks on the inner face, an OMS pod is maneuvered for installation on the aft fuselage. (NASA)

Above right: The OMS pods were detachable so that servicing could be conducted away from the rest of the Orbiter. Shown here is the location on the aft fuselage to which a pod will be attached. (NASA)

Right: An OMS pod in the process of being installed on the Orbiter. (NASA)

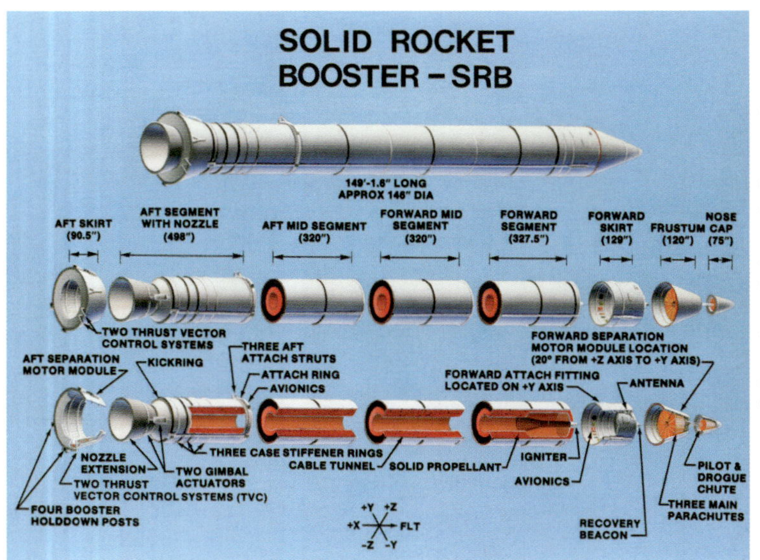

Left: The Solid Rocket Booster (SRB) contained a Solid Rocket Motor (SRM) assembled in segments with nose cap and tail section added. (Thiokol)

Below left: The separate SRB segments were stacked together in the Vehicle Assembly Building at Kennedy Space Center, the outer periphery being where the tang and clevis joint were brought together with O-rings. (NASA)

Below right: An unusual view looking up the length of the mated ET, Orbiter and two SRBs, clearly showing the attachment points. (NASA)

Separated little more than two minutes after launching the Shuttle, the two SRBs parachute down into the Atlantic Ocean. (NASA)

Above: Still full of air and in the vertical position, with the parachutes now removed, the two SRBs are brought together for repositioning into a horizontal mode. (NASA)

Right: The recovery vessel *Freedom Star* arrives back at Port Canaveral towing an SRB. (NASA)

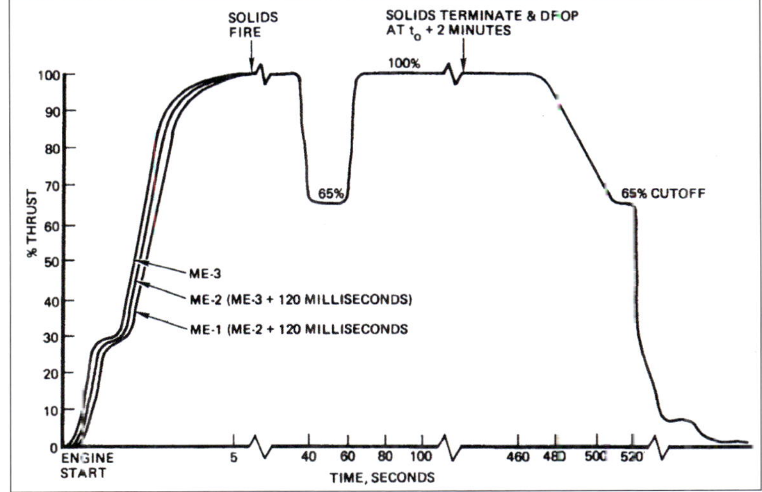

The integrated thrust profile over time from ignition of the SSMEs several seconds before ignition of the SRBs. The thrust profile is that of the Orbiter's main rocket motors and not the SRBs. (NASA)

Chapter 3

Flight Operations 1977–2011

Signed on July 26, 1972, the initial contract for the Shuttle had included a main propulsion test article (MPT-098), a structural test article (STA-099) and two Orbiter vehicles (OV-101 and OV-102) for space operations. It was planned that OV-101 would be ready long before the other Orbiters, and completed to a standard sufficient only for it to fly aerodynamic tests. After those, which would be completed before the first space flight with OV-102, it was to have been refurbished and upgraded for Orbital operations. Highly optimistic, in 1969, NASA spoke of manned orbital flights with the Shuttle beginning in 1975, but by the time the contract was awarded that had drifted to 1977, then quite quickly to 1979.

First, along with engineering tests with MPT-098, NASA prepared for the Air Launched Test (ALT) program in which the 68,000kg (150,000lb) OV-101 Orbiter would be carried on top of a specially converted Boeing 747. Initial flights would verify compatibility before flights in which the Orbiter was released for free flight to a controlled landing. These flights had added importance due to the decision to abandon air-breathing turbofan engines for moving the Orbiters between landing and launch sites and, because of that, the need to validate the captive-carry concept.

The aerodynamic tests of the Orbiter were also necessary for verifying its basic flying characteristics. While a ballistic capsule had a lift/drag ratio of around 0.3, that of the Shuttle at entry was about 1.0 with the angle of descent at 1 degree, allowing considerable margin for going long or short to the required landing site. The maximum L/D in the design was 4.89 but, at 19–24 degrees the approach path in the terminal phase was much steeper than for a conventional aircraft, which typically is 2–3 degrees. On re-entry from space, the Orbiter would be pitched up 40 degrees relative to the horizontal, but that would reduce throughout the terminal descent phase.

The Orbiter's flight control approach was built around a fly-by-wire, digital command system transmitted by data bus using quadruple redundant voting logic to give a high probability that all

Below left: Preparing for space just one lifetime after the Wright Brothers first powered flight of a heavier-than-air flying machine, NASA exuded a confident image of a new era. (NASA)

Below right: STA-099 under construction. During this period of preflight test strategy, the vehicle would undergo ground simulation of flight loads to evaluate structural integrity. (NAR)

NASA boss James Fletcher (extreme left) attends the rollout from Plant 42 at Palmdale, California, on September 17, 1976, of Orbiter *Enterprise*, named after the *Star Trek* spaceship. The event was attended by some of the cast. (NASA)

commands were correct. The pilot's stick input was sensed by four electrical switches, digitized, and sent to the four computers where it would be combined with multiple sensor inputs and the appropriate effector commands issued. These commands would be force-voted hydraulically in the quadruple servo actuators for transmission to the primary actuator. There was no mechanical or analogue backup system.

The ALT flights were designed to evaluate the handling, data input and control functions of the Orbiter, albeit a slimmed-down electronic version with mock-up engines in place of the real thing. Supported by a nose strut and attachment struts at the rear, the first of five captive flights took place on February 15, 1977, followed by three airdrops from August 12. With the nose strut effectively pitching up OV-101 with respect to the datum line of the carrier aircraft, and with the Boeing 747 at full throttle and in a gentle dive, lift was effectively held by the Orbiter. When it dropped the 747 at release, the Orbiter and its carrier turned in an opposing direction to avoid contact.

Three flights were made with a tail cone in place of the simulated engines and two with the tail cone off and the mock-up SSMEs installed, effecting in the latter case the steep descent path that would be characteristic of the terminal descent phase in all Orbiter landings from space. After the aerodynamic tests in 1977, it was decided not to upgrade OV-101 but to incorporate weight-saving improvements and rework STA-099 for orbital flight instead, re-designating it as OV-099 under a contract awarded on January 1, 1979. It would make its first flight on April 4, 1983.

Known as the Mate/De-Mate Facility (MDMF), a special gantry was erected at Edwards Air Force Base for raising the Orbiter off the ground, enabling the Boeing 747 to move into position for receiving *Enterprise* at installed fixtures on the top of the fuselage before backing out. (NASA)

There had always been a desire to have more than two Shuttle Orbiters for space operations, and on January 29, 1979, NASA awarded contracts for OV-103 and OV-104. A fifth Orbiter was planned, and structural assembly of the crew module for OV-105 began on February 15, 1982, ten months after the first Shuttle flight with OV-102. But money was scarce, as NASA's budget continued to decline, and the attempt to get a fifth Orbiter stalled, leaving a lot of structural parts in storage.

All four Orbiters were operational and flying space missions when OV-099 (*Challenger*) was destroyed shortly after launch on January 28, 1986, in full view of spectators at the Kennedy Space Center. Supplementary funds were granted by Congress to complete OV-105 (*Endeavour*) and a contract for that was awarded on July 31, 1987. Following the *Challenger* disaster, the Shuttle program resumed flight operations with the launch of OV-103 (*Discovery*) on September 29, 1988, and *Endeavour* joined the remaining three Orbiters on May 7, 1992.

For almost 11 years, the four Orbiters sustained a vigorous and highly successful program of launching satellites and assembling the International Space Station. But, on February 1, 2003, OV-102 (*Columbia*) was destroyed while returning to Earth after a 15-day mission, and flights did not resume until July 26, 2005. This time there was no appetite for more Orbiters, and the remainder of the flight program operated with the three remaining vehicles – OV-103 (*Discovery*), OV-104 (*Atlantis*) and OV-105 (*Endeavour*), with the final flight of all three on July 8, 2011.

Naming the Orbiters had been a politically sensitive issue. OV-101 had initially attracted the name *Constitution*, but there was a media-savvy shift to *Enterprise*, the name of the *Star Trek* spaceship and the one approved by President Gerald Ford. The remainder were named after famous sailing ships, *Challenger* being a British exploration vessel and *Endeavour* the name of the ship assigned to the British explorer Captain Cook, which is why it bears the British-English spelling.

Primary landing sites for the Orbiter were the concrete runways at the Kennedy Space Center (15 or 33) or Edwards AFB, California, (05/23, 15/33 or 17/35) with an emergency landing field at White Sands, New Mexico, (17) in case of flooding at Edwards. Mission profiles ran on the intact-abort concept, with a plan for an Orbiter incapable of reaching orbit returning to KSC, or of crossing the Atlantic Ocean with landing options at Naval Air Station Bermuda, Lajes Air Base (AB) in the Azores, Zaragoza AB in Spain, Morón AB in Spain or Istres AB in France. There was also provision for landing at RAF Fairford in the UK, but that would only apply to the most northerly of orbital inclinations.

Below left: The first of eight captive-carry flights with *Enterprise* on the Boeing 747 was conducted on February 18, 1977, during which free-flight crews were on the Orbiter flight deck. (NASA)

Below right: The first free-flight begins as *Enterprise* "drops" the Boeing 747 carrier plane at an altitude of 7,346m (24,100ft) and starts down on a descent lasting 5mins 22secs with a rollout of less than 3,353m (11,000ft). (NASA)

Above: With the tail cone attached for less drag, each free-flight descent lasted about the same time as the first, and several banks and turns were made during descent to evaluate flying characteristics. (NASA)

Right: This image shows dummy main engines being installed in *Enterprise* for flights without the tail cone for better simulation of the true flying qualities of the Orbiter in its operational configuration. (NASA)

Below: The first of two flights with the tail cone removed took place on October 12, 1977, the Orbiter dropping steeply and landing on the ground in 2mins 35secs. (NASA)

Under original plans to fly polar-orbit missions from Vandenberg AFB, California, Orbiters would have touched down on the modified runway at that facility. The first launch from Vandenberg (STS-62A) was scheduled for July 1986, before the *Challenger* loss in the preceding January brought a halt to all flights and completely cancelled all further plans for launches from the West Coast.

Phase One

In retrospect, it is possible to divide the Shuttle program into three distinct phases beginning with the first flight, although these are not official and are rather based on a personal and subjective view of how the program evolved through a wide range of changing circumstances. The first phase saw the planned use of the vehicle and its payload capabilities through adaptation to changing economic circumstances of NASA and the United States.

As explored in Chapter 1, a high annual flight rate was key to the economic goal of the reusable Shuttle, which was to support a space program that, over a defined period, would be cheaper than one based on expendable rockets and manned space capsules. The desire to have a Space Tug to move satellites to and from higher orbits than the Shuttle alone could reach was sacrificed on cost grounds. But there was an alternative, one which appeared to achieve some of what the Tug would have provided.

The 1970s saw a major expansion of commercial space activity, with many countries around the world wanting satellites for communication, TV, weather forecasting, and environmental resource monitoring. This was the age of the environmentalists, their arguments for a cleaner planet based on the Middle East crises and the burgeoning cost of oil, which began a global recession. From the White House down, from 1977 to 1981, the Carter administration pushed for global environmental monitoring, reduction of hydrocarbon fuels, and ozone layer-busting chemicals, and America led the way in banning ozone-depleting fluorocarbons.

NASA was charged with supporting this, and the first great wave of commercial space activity saw many countries buying satellites from American companies and having them launched on what NASA called "reimbursable" deals, in which they paid for only the cost of getting them into space. It quickly became apparent that NASA could significantly increase the Shuttle flight rate if all expendable rockets were retired, and these national and international customers were moved to the Shuttle. The Orbiter could carry two or three satellites into low Earth orbit, each attached to a commercially developed booster rocket that moved each to its desired orbit.

Developed by McDonnell Douglas, it was known as the Payload Assist Module (PAM) and came in three sizes, capable of lifting (22,300 miles) satellites weighing up to 1,250kg (2,750lb) for PAM-D, 1,880kg (4,100lb) for PAM-DII and 1,996kg (4,400lb) for PAM-A to geosynchronous orbit at an altitude of 35,900km. The letter suffixes denoted that they could also be used for Delta (D) and Atlas (A) expendable launch vehicles. This provided interchangeability during the period when conventional rockets were still launching satellites. Geosynchronous altitude was the operating region for many application (weather, TV broadcast, communication, navigation, etc.) satellites, a location above the Earth's equator where they would appear to remain stationary because their orbital period was the same as the Earth's speed of rotation. These barrel-shaped PAM rocket motors, when attached to the bottom of a satellite, could propel it to the required position, and in the case of the Shuttle would be an integral part of each satellite, as it was supported on its dedicated cradle in the Orbiter payload bay. When deployed, the cradle would spin up the satellite and its PAM motor, so that when it was released vertically up and out of the bay, it would be spin-stabilized until the PAM motor was fired after the payload and the Orbiter had drifted a safe distance away.

Right: The pilots assigned to the five Approach and Landing Test (ALT) flights conducted between August 12 and October 26, 1977: Gordon Fullerton and Fred Haise (crew 1), and Joe Engle and Richard Truly (crew 2). Crew 1 flew three free-flight tests and crew 2 flew two. (NASA)

Below: The first Shuttle to reach the launchpad was *Columbia*, although it was delayed by problems with the thermal protection system and other technical issues. (NASA)

Columbia commander and veteran astronaut John Young (left) and Robert "Bob" Crippen at breakfast on the morning of launch from the Kennedy Space Center. (NASA)

A more powerful booster stage known as the Inertial Upper Stage (IUS) was developed, a much more powerful two-stage motor which could push a 4,940kg (10,890lb) satellite to geosynchronous transfer orbit. Developed by Boeing for the Air Force in the mid-1970s, it was used for government payloads by both NASA and the Department of Defense but never for commercial satellites. Unlike the Tug, all these boost motors had a single use and could not perform fetch-and-carry missions. However, they at least allowed the Shuttle to enter the launch vehicle market for commercial as well as government payloads.

Great hopes were pinned on the first Shuttle launch on April 12, 1981, exactly 20 years after the historic flight of Russia's Yuri Gagarin, the first person to orbit the Earth. The option for sending an unmanned Orbiter into space before humans were aboard was never an option, the complexity of adapting it for such a task negating justification for doing that. In what has been described as the highest-risk flight in manned space operations, veteran astronaut and Moonwalker John W. Young and newbie astronaut Robert L. Crippen took *Columbia* on a two-day flight before landing at Edwards AFB. It was followed by three more demonstration flights, the last in June 1982 where President Ronald Reagan welcomed the crew back home.

After the first four Shuttle flights, NASA boasted that it was declaring the Shuttle an operational space transportation system, all flights carrying the initial letters "STS." The ALT flights and the first four orbital flights had two-man crews and ejection seats with break-through panels above their respective positions on the flight deck. From STS-5, however, NASA declared the system open for business and the job of recruiting customers for an increasingly busy flight manifest attracted commercial clients from several foreign countries.

On the first of those launched in November 1982, STS-5 carried a communications satellite for Canada and another for a commercial telecommunications company in the US. A year later, *Challenger* began flight operations and launched a tracking and data relay satellite to geosynchronous orbit to provide continuous communications between Earth and the Shuttle, as well as an increasing number of unmanned satellites and spacecraft as more were added to the constellation in orbit.

The initial operational phase of the Shuttle program was built around this satellite-launching capability, while behind the scenes NASA began to think once more about the space station for which the Shuttle had been intended all along. In the meantime, *Discovery* joined the fleet in 1984, followed by *Atlantis* the following year. In November 1983, *Columbia* carried the European Spacelab research laboratory into orbit, product of a cooperative venture with the ESA. Contained within the payload bay and accessed via a hatch in the Orbiter mid-deck and a short tunnel, it provided a research opportunity for the crew of six, including the German astronaut Ulf Merbold. Many more Spacelab flights would follow.

By mid-January 1986, 24 Shuttle flights had been completed, and there was high optimism that the various expendable launch vehicles that had been sending satellites and spacecraft into orbit would soon

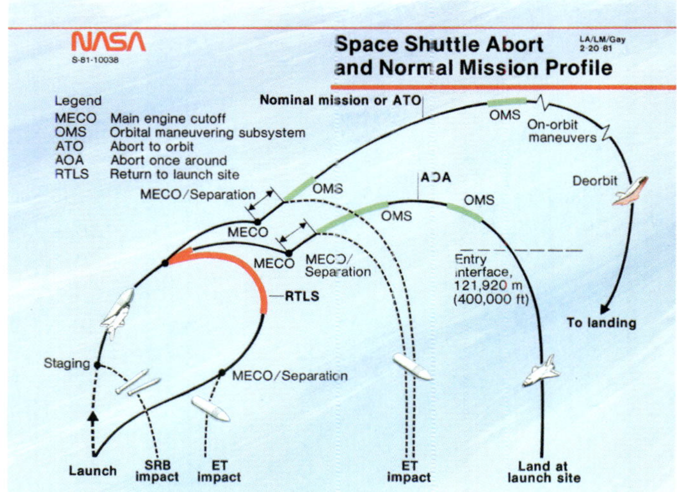

Above left: A descriptive chart showing the trajectory and event sequence for a normal mission and the abort options for safe recovery after a malfunction during ascent. The Abort Once Around (AOA) option places the Orbiter on a trajectory once around the Earth for a landing back at the Kennedy Space Center, which is also the landing site for the Return To Launch Site (RTLS) option. (NASA)

Above right: *Columbia* lifts off at midday on Sunday April 12, 1981, 20 years after the first human – Yuri Gagarin – made the first flight around the Earth. (NASA)

be retired. To accommodate an increasingly complex arrangement of payloads on the Shuttle manifest, and to assign a specific mission number to a specific set of payloads, a new mission numbering sequence was introduced after STS-9, the first Spacelab mission. It would also serve to track a particular mission as it progressed over the months and, in some cases, years between assignment and launch.

It was created as a means of identifying when the flight was to take place, the site from which it was launched and the sequence in that year when it was assigned to fly. For instance, the first mission after STS-9 was STS-41B, the "4" indicating fiscal year 1984, the "1" indicating KSC (rather than Vandenberg, which was "2") and the "B," the second flight in that year. No flights ever took place from Vandenberg, so the "2" series, while being assigned for planning upcoming missions, was never flown.

On the last flight before *Challenger* was destroyed, Congressman Bill Nelson rode *Columbia* into space for a six-day flight when the seven-man crew, including astronaut Charles F. Bolden, placed a communications satellite into orbit. Bolden would fly three more missions, the last of which he commanded, before serving as the boss of NASA from 2009 to 2017. Nelson has served in that capacity since May 2021.

The first phase of the Shuttle program ended on January 28, 1986, when, after launching in freezing temperatures, the O-rings ostensibly protecting the tang and clevis joints from leaking hot gases between SRB segments failed. The ensuing blast of hot gas severed an aft SRB restraining strut causing the booster to pivot around the forward strut and break open the ET, igniting the cryogenic propellants in the ET and separating *Challenger* and the two SRBs in the ensuing blast.

The catastrophe occurred just 73 seconds after lift-off, in full view of crew families, VIPs, visitors, guests, and NASA personnel at KSC. Among the seven-person crew was teacher Crista McAuliffe, and the launch was beamed live to school audiences across the country in preparation for live lessons she was to have conducted from orbit. The event brought a state of national grief and an awakening of just how dangerous and fraught the business of manned space flight was, ultimately changing the orientation of the entire US space program for ever.

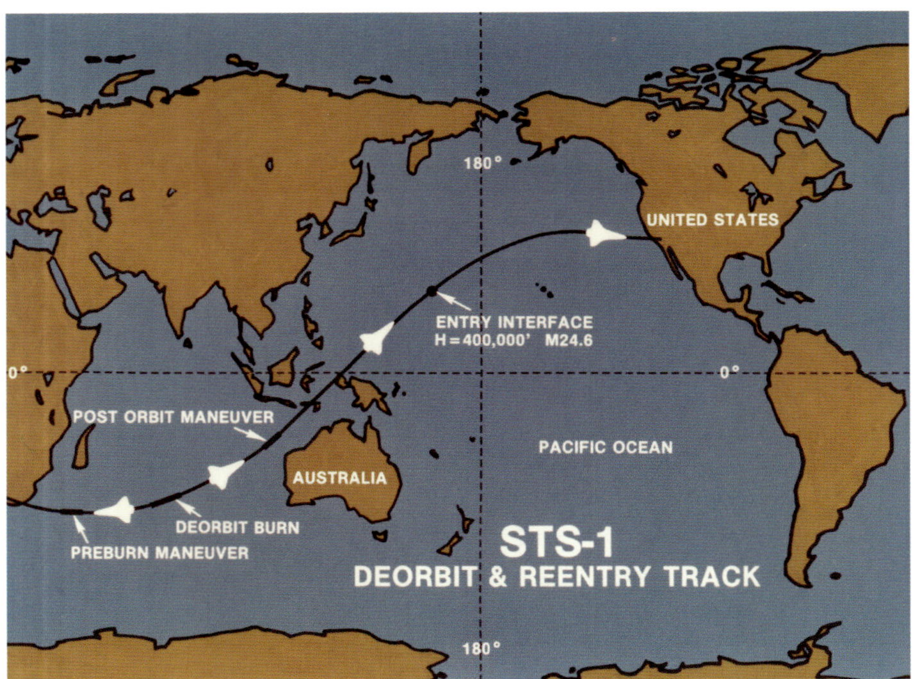

Left: Getting the Shuttle Orbiter back on the ground begins around the other side of the world, with a deorbit burn to begin the long descent toward the atmosphere. (NASA)

Below: This image shows the terminal phase of descent, together with the approach and landing phase, at the then-Dryden Flight Research Center, now known as the Armstrong Flight Research Center. (NASA)

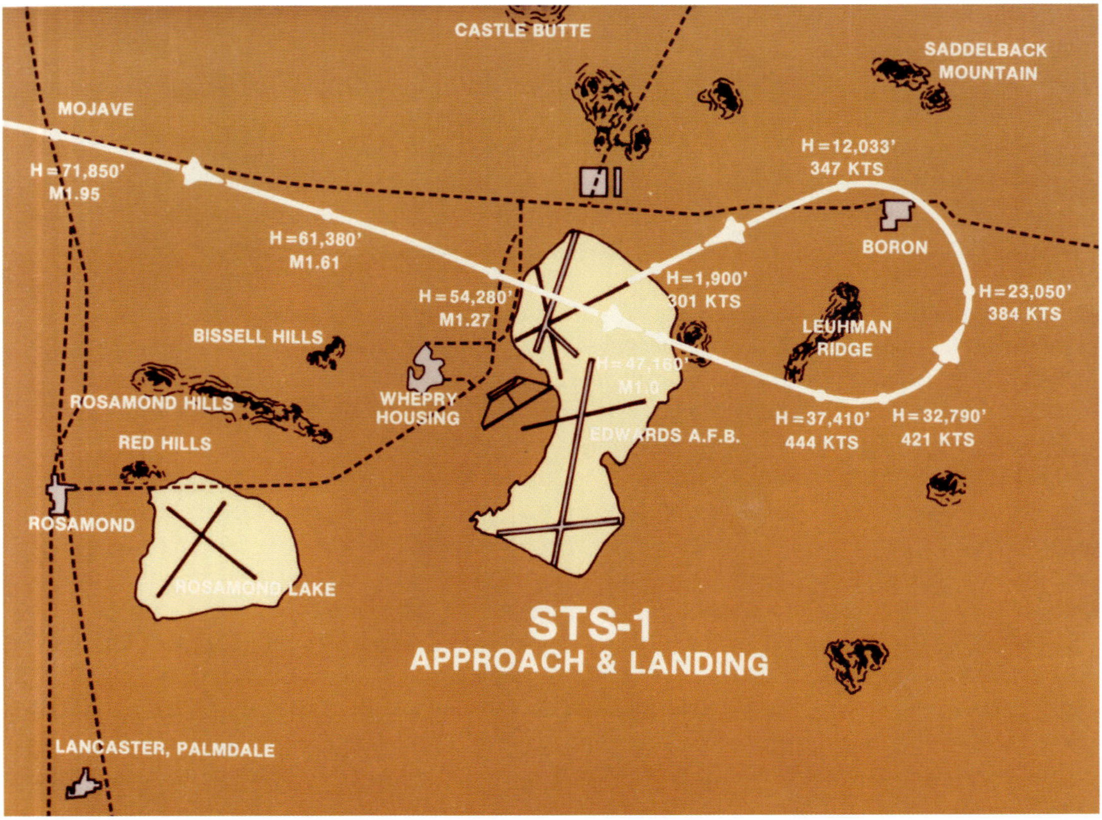

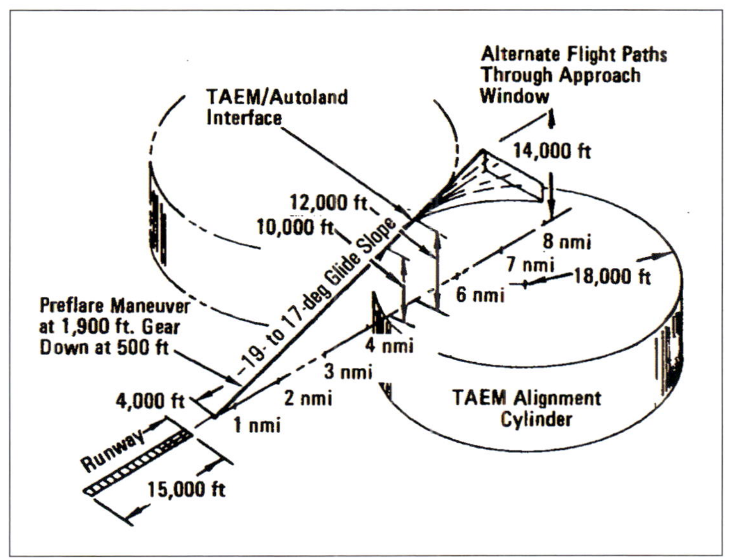

Above left: Two Terminal Area Energy Management "cylinders" are available for defining the heading alignment during descent, the degree of flight around either is dependent on the angle at which it is intercepted. (NAR)

Above right: Little more than two years after the first Shuttle flight, on June 18, 1983, NASA launched Sally Ride, the first American woman to reach space, aboard *Challenger* on STS-7. Seen here in orbit, Ride poses near the Electrophoresis Operation in Space (EOS) experiment, a pharmaceutical research project. (NASA)

Below: With several successful flights accomplished, on November 28, 1983, *Columbia* carried the European Spacelab laboratory module so that its six-man crew could conduct experiments in space. (NASA)

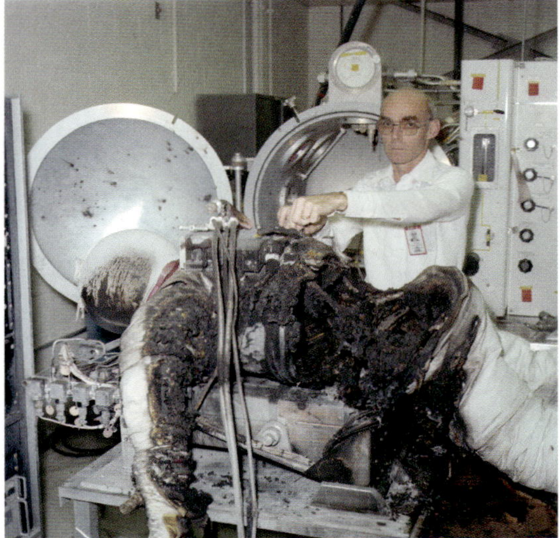

Above left: The first tether-free spacewalker, Bruce McCandless drifts independent of *Challenger* during STS-41B, which was launched on February 3, 1984. (NASA)

Above right: Regarded as the smallest spacecraft ever built, the Extravehicular Mobility Unit (EMU) made spacewalks possible, but it also had its own development problems. Here, a manikin has melted in an accident during 1980, resulting in a fire. (NASA)

Above left: Launched on April 6, 1984, *Challenger* left a long duration platform in orbit for retrieval on a later mission. It also conducted a repair job on the Solar Maximum Mission (SMM) satellite, shown here attached to the Orbiter's robotic arm. (NASA)

Above right: STS-51A (*Discovery*) sent five astronauts into space on November 9, 1984, to retrieve two commercial satellites stranded in orbit, one depicted here, for refurbishment and resale. (NASA)

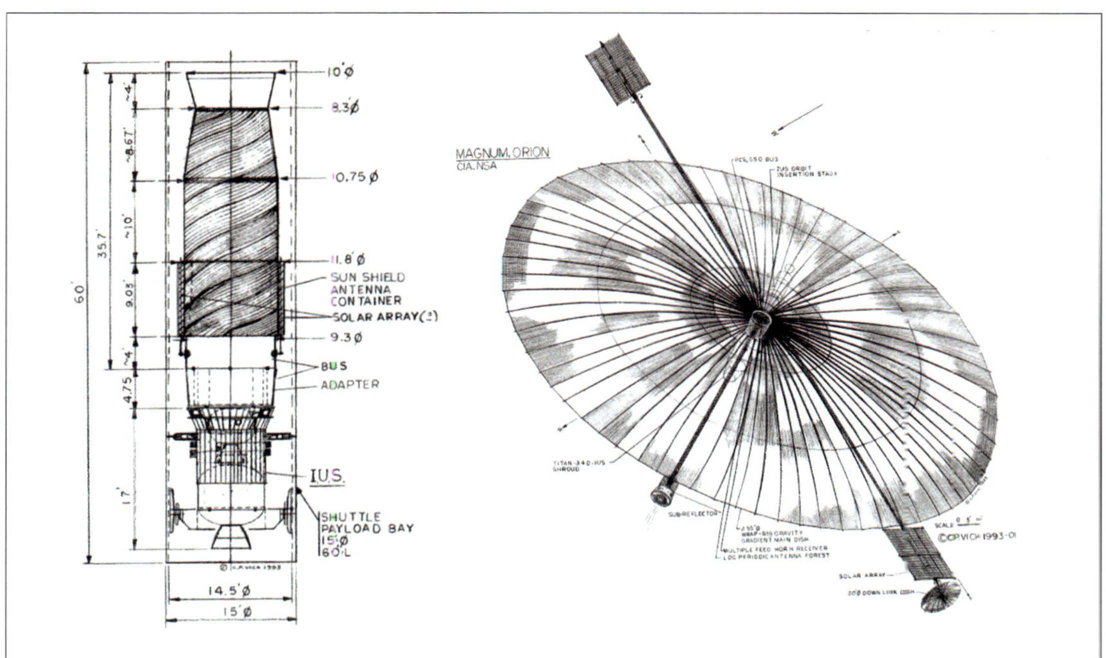

Above: On January 24, 1985, STS-51C carried the first substantial US military payload when *Discovery* lifted a still-secret payload believed to be the Magnum signals intelligence spacecraft operated by the National Reconnaissance Office for the CIA, a speculative drawing of which is shown here, with the IUS attached for lifting it to its operating orbit. (Author's collection)

Below left: Operating as a delivery truck, the Orbiter could carry up to three satellites to be released into orbit or a suite of experiments for operation in the vacuum of space, as shown here in the payload bay of *Discovery*, launched June 17, 1985, on STS-51G. (NASA)

Below right: After launch on August 27, 1985, *Discovery* astronauts on STS-51L rescued a communications satellite stranded in low Earth orbit. (NASA)

Above left: The launch of STS-51L on January 28, 1986, showing the puff of black smoke on the lower segment of the right-hand SRB, which would cause hot gases to sever the attachment arm, causing the booster to rotate around the upper arm and split open the top of the ET, causing destruction of *Challenger*. (NASA)

Above right: The payload carried by *Challenger* when it was destroyed was a Tracking and Data Relay Satellite (TDRS), designed to provide continuous communication between the Shuttle and satellites and ground stations. The first in a series of TDRSs was launched by STS-6 in 1982, and is shown here with the IUS attached. (NASA)

Phase Two

The Shuttle was grounded for 32 months following this incident, during which a presidential commission investigated many aspects of the entire program. Plans to fly from Vandenberg on defense-related missions were abandoned, as was the entire concept of flying US commercial and foreign, non-government satellites. Inherently reluctant to place all its payloads on the Shuttle, the USAF had already reinvigorated the flagging expendable launch vehicle market, which also saw a return to production to launch the commercial satellites no longer assigned to the Shuttle.

Back into full-scale production went the Delta launchers the Shuttle had been expected to replace. Atlas and Titan had a more assured future due to the USAF maintaining those as backup to the Shuttle program, and many procedures and processes were put in place to ensure greater safety margins. There was now none of the "launch on time – every time" mentality, which had pushed the entire program to dangerous levels with inadequate resources to cope with unachievable goals demanded by the drive to achieve impossible flight rates.

Technical changes too were made in the design of the SRB segment joints, with greater protection from failure. Ice-teams would be sent out before each flight to alert mission controllers to potential threats from cold temperatures. There had been plans to launch planetary spacecraft by carrying them up attached to a cryogenic Centaur stage inside the payload bay. This had potential risk because of the explosive nature of hydrogen and oxygen when ignited, and the idea was dropped, replaced on those missions by the two-stage IUS. The consequences lasted many years after flight resumed, as science experiments, satellites, and payloads queued to get a ride on an expendable launcher. The USAF also had payloads that could only fly on the Shuttle, and those were added to the manifest over time.

Mission assignments dispensed with the complicated designations referred to earlier and returned to a simple numerical sequence of specific payloads, but they did not always fly in sequence, STS-28 for instance flying after STS-29 and STS-30. But the number meant a specific payload manifested for a particular flight, which was a much simpler system of designating missions by payload.

The second phase of Shuttle operations began with the return-to-flight on September 29, 1988, and lasted until 1998, during which time a wide range of government payloads, satellites, astronomical observatories, and Earth resource payloads were carried into space. Delayed by the restoration of flight operations and the funds required to build a replacement for *Challenger*, plans for a space station had already been set in motion in 1984 when NASA Administrator Jim Beggs received approval to sign up international partners from Europe, Japan and Canada.

Initially called *Freedom*, the station would be assembled by the Shuttle, with modules contributed by the ESA and Japan and the robotics provided by Canada, which had already provided remote manipulators for the Orbiters. Gone were the days when NASA wanted to launch a fully equipped station on a Saturn V rocket, capable of supporting 12 people. Instead, a modular space station would be put together in orbit, and for that the Shuttle was the perfect truck. After assembly, it would support the station through crew exchanges and the provision of food and general logistics, as well as carrying experiments up and down, and adding additional elements as required.

Above left and above right: The Magellan spacecraft launched by Atlantis on May 4, 1989, inside the payload bay (left) and in space during deployment with the IUS booster stage attached. (NASA)

The early 1990s saw the collapse of the Soviet Union and the emergence of Russia and associated territories as federated states. Since the 1970s, the Russians had pursued development of their own space stations, called *Salyut* and *Mir*, together with manned Soyuz crew capsules and Progress logistics craft. They had chosen this path of building space stations after the success of NASA's Apollo program and a succession of four failures in four attempts with their Moon-booster called N-1.

Since *Freedom* was first proposed by the White House in 1984, NASA had waited for nearly a decade for the money it needed from Congress to start building it. Each year, Congress came close to cancelling the project, and when Bill Clinton became President in January 1993, he ordered a full review. In a similar position, the Russians had scant resources to build on previous successes, and to begin assembling their own *Mir-2* station. Seeing an opportunity to share resources and expand the potential, on September 2, 1993, both sides signed a partnership agreement for what was now to be called the International Space Station.

There was much for the Americans to learn from the Russians. In turn, the Russians received support for what was potentially an unsustainable program of national expenditure at a time when the country faced economic pressure from its newly found democracy. NASA received a more favorable hearing in Congress, with Russia providing two major core modules with which the ISS would be assembled. Slowly, confidence began to build back in what was envisaged as the biggest international engineering venture of all time.

An artist depicts the Galileo probe attached to its IUS stage after separating from *Atlantis* following launch on October 18, 1989. It began operation in orbit around Jupiter six years later, deploying a probe through the atmosphere and sending back data for almost 14 years. (NASA)

Above: Deployed by *Discovery* after launch on April 24, 1990, the Hubble Space Telescope was revisited several times, being improved and upgraded over five servicing missions between 1993 and 2009. It is cleared to remain in service until 2026. (NASA)

Right: The first job for *Endeavour* when it entered service on May 7, 1992, was to fly a rescue mission for an Intelsat communications satellite stranded in low orbit. Manhandled by three astronauts, it received a new boost and was set free to depart for its intended orbit. (NASA)

 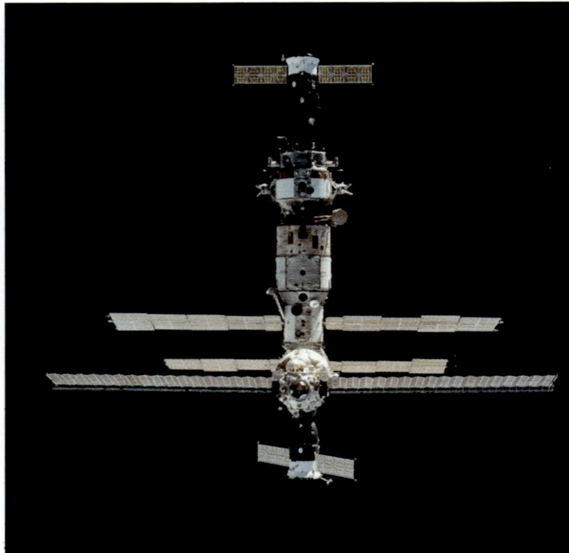

Above left: *Endeavour* also supported the first Hubble servicing mission following its launch on December 2, 1993. (NASA)

Above right: Following launch on February 3, 1995, *Discovery* rendezvoused with Russia's *Mir* space station in the first of a series of missions following agreement that Russia would join the International Space Station, which was planned for assembly later that decade. (NASA)

Below: This is the payload bay of *Atlantis* as it nears the *Mir* space station on June 29, 1995, two days after launch, displaying the extended Spacelab module, and the new docking module. (NASA)

All this gave the Shuttle renewed purpose, but, in February 1995, before assembly of the first elements could begin, NASA flew a rendezvous mission to Russia's *Mir* station followed by nine docking flights between June 1995 and June 1998. On those missions, cosmonauts and astronauts learned to live and work together in space and develop working practices as a prelude to building the ISS. This relationship would last for several decades and is a story best told elsewhere, but insofar as the Shuttle was concerned, it has finally been given the opportunity to fulfil its intended purpose back in the late 1960s.

Right: Russia's three departing *Mir* crew took this shot of the docked space vehicles from their Soyuz spacecraft on July 4, 1995, leaving two Russians aboard *Mir*. *Atlantis* returned with the eight planned crew members, the greatest number to date on a single Shuttle return from orbit. (NASA)

Below: A close-up view of the third Shuttle docking on *Mir*, achieved by *Atlantis* on March 24, 1996. (NASA)

Launched on January 12, 1997, STS-81 was the fifth of nine missions to *Mir* by *Atlantis*, seen here through the docking alignment calibration index on the Orbiter. (NASA)

Astronaut Andrew Thomas signs the visitors' book aboard *Mir* during the last visit by Shuttle to the Russian station, following the launch of Discovery on June 2, 1998. (NASA)

Phase Three

The first ISS element launched was Russia's *Zarya* module, lifted to orbit by a Proton rocket in November 1998. It was followed by NASA's Unity module carried by *Endeavour* and docked to *Zarya* to begin the full assembly, a sequence not completed for 13 years. We can consider the launch of Unity as the beginning of the third and final phase in Shuttle flight operations, following 67 flights after the *Challenger* disaster.

This third phase is characterized by fully operational support for the ISS until the Shuttle was retired in July 2011 after 43 flights. It saw several historic events, including the establishment of the first permanent occupation of the ISS, still in its initial configuration and far from complete, on November 2, 2000. The ISS, now complete, although frequently updated with new equipment, scientific instruments, and operational systems, has been permanently manned from that day to this and is expected to remain so until 2030.

Throughout this final phase, NASA continued to use the Shuttle to carry out major launches, including the Hubble Space Telescope in 1990, which was revisited five times between 1993 and 2009, a capability impossible without the reusable transportation system. It also launched the Chandra X-ray observatory and the Shuttle Radar Topography Mission to provide for the international community of geophysicists; the most complete topographical map of the world ever made.

However, just as the *Challenger* disaster in 1986 marked a dramatic shift in Shuttle utilization, the loss of *Columbia* on February 1, 2003, triggered the decision to retire the Shuttle and once again set NASA on a very different path to the future. The tragedy came about when a piece of spray-on foam insulation broke away from the ET after launch and struck the brittle RCC thermal protection on the

More than 36 years after becoming the first American astronaut to orbit the Earth on February 20, 1962, John Glenn went back to space aboard *Discovery* on October 29, 1998, the 77-year-old veteran spending over 19 hours in orbit. (NASA)

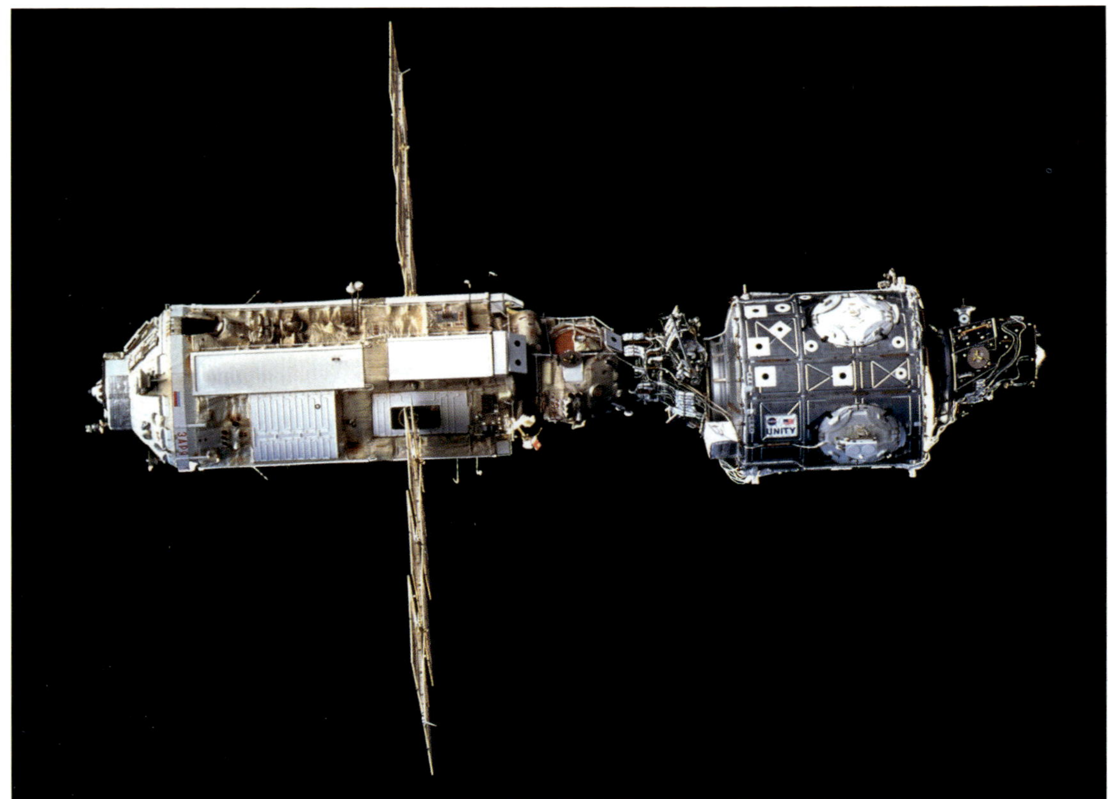

Launched by Proton rocket in November 1998, Russia's Zarya module (left) received NASA's Unity module (right) delivered by *Endeavour* after launch on 4 December, beginning the long process of assembling the International Space Station (ISS). (NASA)

wing leading edge. It shattered the RCC panel and opened a cavity, through which hot gases during descent melted the wing, and eventually the Orbiter, from the inside. All seven crew members were lost as debris from the vehicle fell across several US States.

This time, the Shuttle was grounded for 30 months, during which time a procedure was established for using the Orbiter's robotic arm for conducting a video survey in orbit of all the vulnerable areas with the possibility of using a special repair kit for addressing any damage. It always had only a moderate prospect of success, but as a precaution all flights were assigned as ISS delivery missions so that the crew could wait it out aboard the station for a rescue flight using another Shuttle Orbiter. There was one exception. Permission was given to conduct the last Hubble servicing mission, which meant the Orbiter could not reach the station in the event of an emergency. It was successfully carried out by *Atlantis* in May 2009, much to everyone's relief.

By the time the last Shuttle, *Atlantis*, came home on the morning of Thursday July 21, 2011, the future for NASA looked very different. America no longer had any capacity for sending people into space. Crew members to the ISS would now fly on Russian Soyuz capsules and logistical provisions would be uplifted by Progress vehicles, but only until America could begin using commercial spacecraft developed by private companies. The decision to end Shuttle operations was based only in part on the shock of losing a second Orbiter in 2003, partly on grounds of cost and partly as a redirection of effort for the future.

Above: On February 11, 2000, *Endeavour* began an 11-day extended solo mission to conduct topographic measurements of the globe, providing vital information from a synthetic aperture radar carried on a long boom extended from the payload bay. (NASA)

Right: Not all later Shuttle flights were for the ISS. On July 23, 1999, *Columbia* carried the Chandra X-ray observatory into orbit, complete with its IUS boost motor. It is still operational. (NASA)

Highly detailed analysis of the risk of losing an Orbiter and its crew revealed that there had always been a high probability of failure somewhere among the myriad systems and subsystems essential for crew safety. Initially, and before it started flying the Shuttle, NASA had thrown around safety figures predicting a catastrophic loss of one in one-thousand flights. In reality, there had been two in 135. Analysis showed that a range of modifications and upgrades would significantly improve the safety record, but expenditure there would starve the agency of funds to invest in future manned spaceflight goals.

Recognizing all these factors, during the decision period after the loss of *Columbia*, NASA considered new solid propellant boosters, even replacing the SRBs with liquid propellant boosters of greater thrust and with improved safety factors. Recovery would have been problematic due to the vulnerable components of a liquid rocket motor, but some thought was given to rotating clamshell doors, which could have closed during descent, providing a watertight enclosure to protect them.

But there were calls for NASA to get back into doing what it had traditionally done best: conducting deep-space exploration. The decision was made to retire the Shuttle and assign those funds to a long-term program aimed at exploring the asteroids and beginning a transition to permanent research bases on the Moon and eventually on Mars. None of that was possible while also funding the Shuttle.

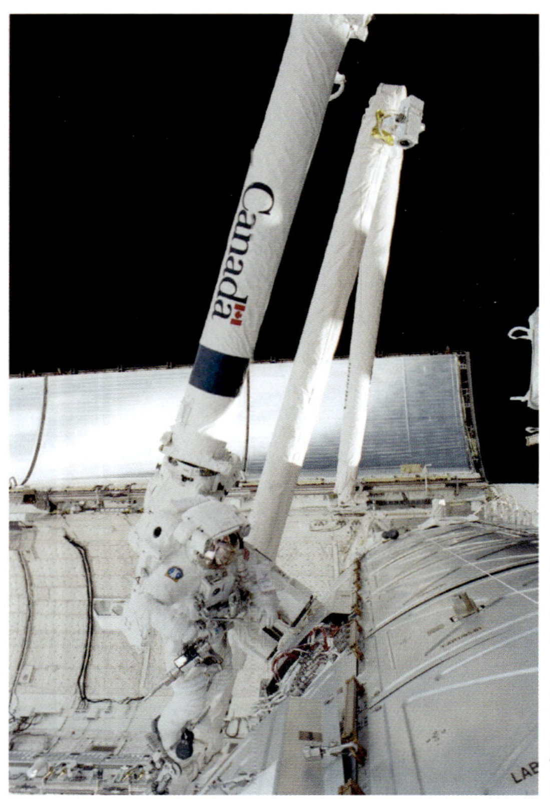

Toward that end, Space-X and Boeing separately took on the task of providing capsules to carry astronauts back and forth to the ISS. NASA would concentrate on development of a rocket bigger than the Saturn V that took men to the Moon and spacecraft to send humans back to the lunar surface.

It all took a lot longer – and much more money than anticipated – and commercial flights to the ISS with Space-X Dragon capsules only began in May 2020. For almost nine years, anyone visiting the ISS had to fly in a Russian Soyuz capsule. Although delayed, the commercial program eventually came to fruition, but NASA's plan to build a super-booster and send people back to the

Left: *Endeavour* carried the European *Raffaello* logistics module and a sophisticated robotic arm to the ISS on April 19, 2001, the latter to help with assembly of additional elements. (NASA)

Below: Throughout the Shuttle program, great improvements were made to equipment, operating procedures, and personal hygiene, including provision for a toilet, with basic operating principles shown here. (NASA)

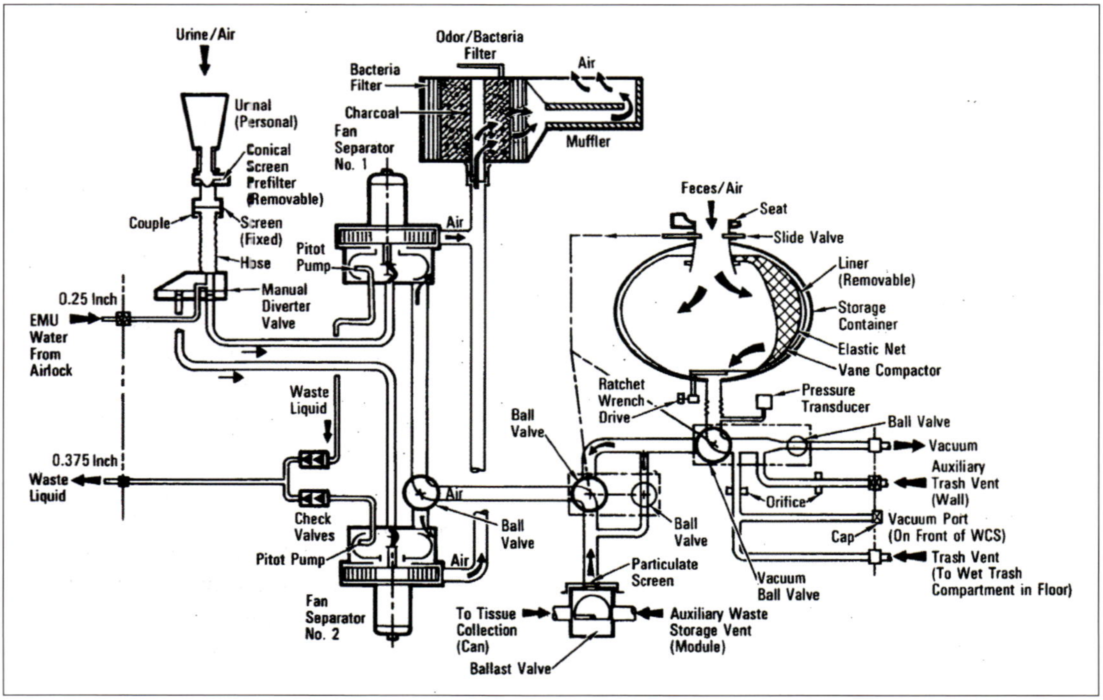

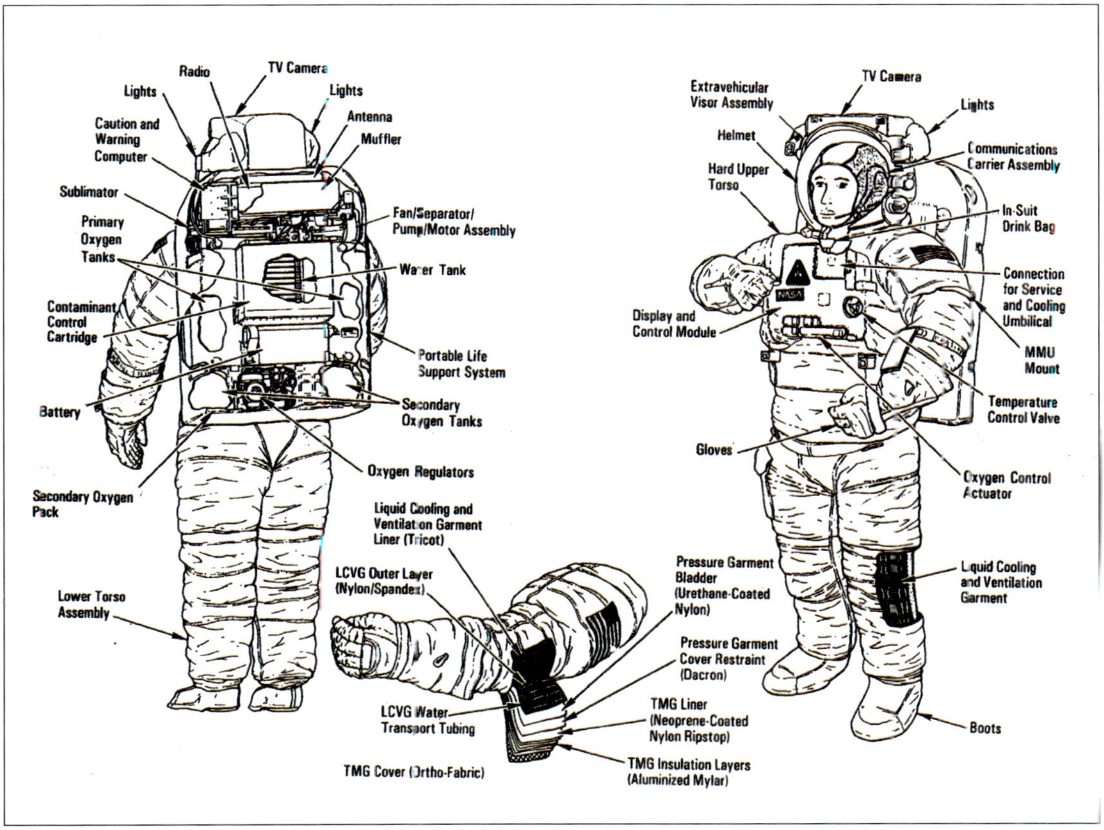

A vital part of Shuttle operations was the ability to work outside, assembling the ISS, repairing elements and satellites, and to conduct visual inspections. The standard IMU has remained unmodified throughout. (NASA)

Launched on June 5, 2002, *Endeavour* got the job of lifting Canada's Mobile Base System to the ISS, a remotely controlled manipulator-arm "truck" running along the exterior of the ISS truss assembly to carry out tasks on the exterior. (NASA)

Moon has run into constant delay, minor technical problems and cost escalation far beyond the funds originally estimated as necessary to retrace the steps of Apollo.

The operational legacy of the Shuttle is pegged also to the vast amount of data about engineering and the ability of humans to remain in space for long periods, which these two programs delivered. As well as sending people to the Moon on a regular basis under what is called the Artemis program, with the exception of Russia, the ISS partners are supporting a space station for lunar orbit. Called Gateway, it will also provide a base for operations to and from the surface of the Moon. Without the experience available from 30 years of flying the Shuttle and more than 20 years of the permanently manned ISS, that would be very difficult to accomplish.

Left: *Endeavour* **carried an essential truss structure to the ISS on November 23, 2002, when astronauts John Herrington (right) and Michael López-Alegria worked on installing cables and connectors. This was the last flight before the loss of** *Columbia*. **(NASA)**

Below: **Downlinked (images sent to the ground) from** *Columbia*, **three crew members of STS-107 – (from top) Laura B. Clark, Rick D. Husband, and Kalpana Chawla – are seen in mid-deck sleep cubicles during their science mission, which lasted almost 16 days. (NASA)**

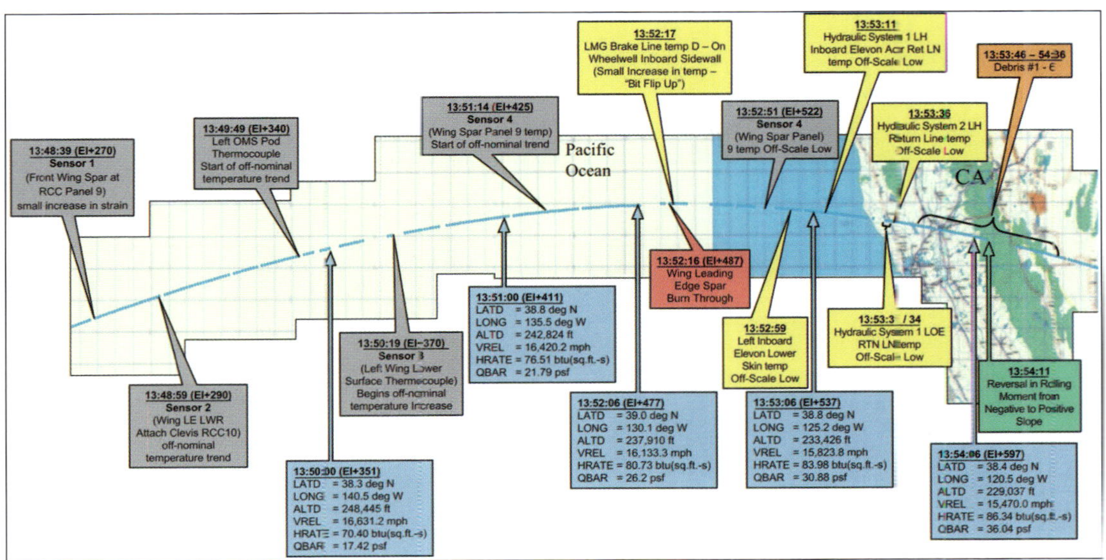

The timeline of critical events associated with the loss of *Columbia* as it broke apart on re-entry, on February 1, 2003. Underscored in each box is the time in UTC (GMT) and in some cases the elapsed time since entry interface (EI). (NASA)

Launched on December 9, 2006, *Discovery* cuts a dramatic departure from LC-39B at the Kennedy Space Center, beginning a 12-day servicing mission to the ISS. (NASA)

Above: When, on July 26, 2005, *Discovery* put the Shuttle program back on track after the loss of *Columbia*, it took along an extension to the Canadian robotic arm for a series of "selfies," which, when stitched together, verified neither the tiles nor the RCC had any damage. (NASA)

Left: The Orbiter and the ET were vulnerable to adverse weather, as shown here with damage on the ET for STS-117 caused by a hailstorm on February 26, 2007, which delayed launch until June 8 while repairs were made. (NASA)

A "selfie" in orbit during STS-117 showed damage to the thermal protection on *Atlantis,* which was not in a critical area for safe return. Other damage was also noted elsewhere. (NASA)

Endeavour (left) rolled out to LC-39B on April 17, 2009, ready to stand by if needed to rescue the *Atlantis* crew on LC-39A (right) prior to its flight to conduct the last Hubble servicing mission beginning May 11. (NASA)

Doug Hurley aboard *Endeavour* on the STS-127 mission launched July 15, 2009, seen in the aft flight deck against the two aft-facing windows and the two windows in the top of the crew compartment. (NASA)

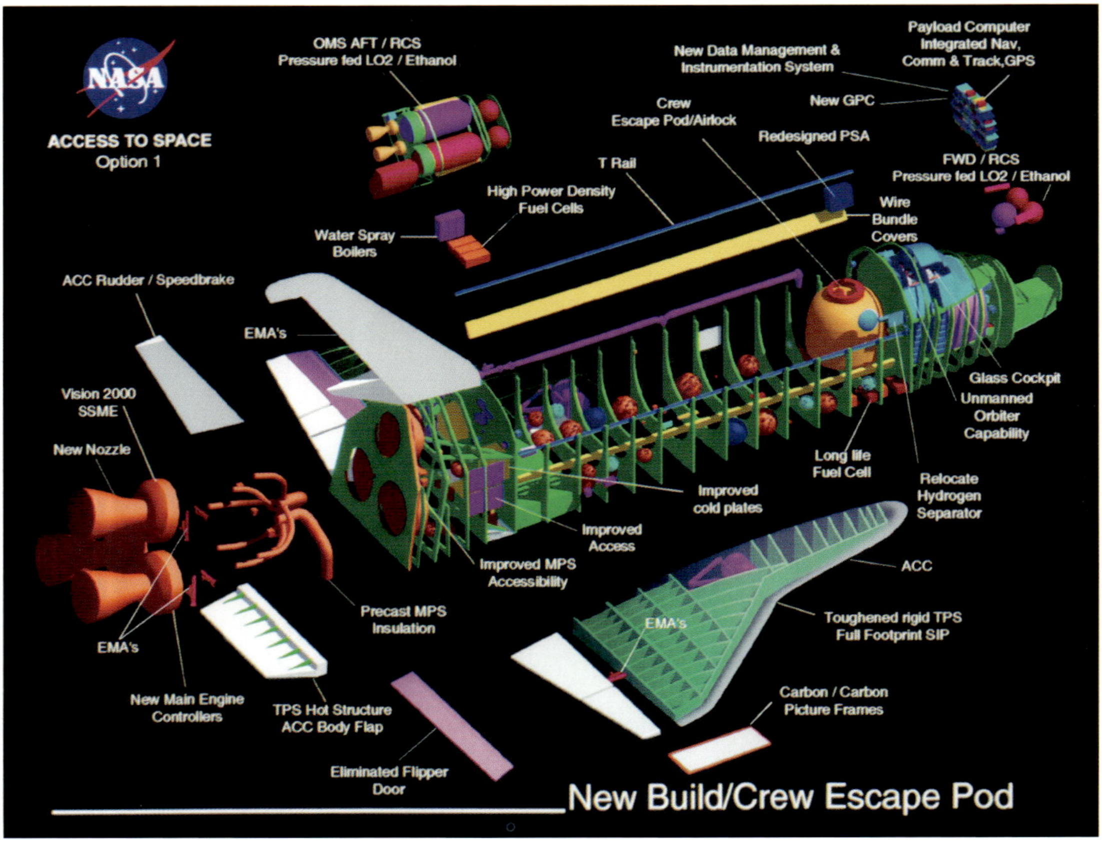

As early as 1993, NASA reviewed potential upgrades and improvements as displayed on this graphic, including the possibility of an escape pod to provide a second means of returning to Earth. (NASA)

Right: Several options were available for using Shuttle elements to lift heavy cargo into orbit, including replacing the Orbiter with a payload pod but at the cost of losing the main engines on every flight. (NASA)

Below: A proposed heavy-lift vehicle carrier could have placed up to 82,000kg into low Earth orbit, almost three times that of the Shuttle/Orbiter combination. (NASA)

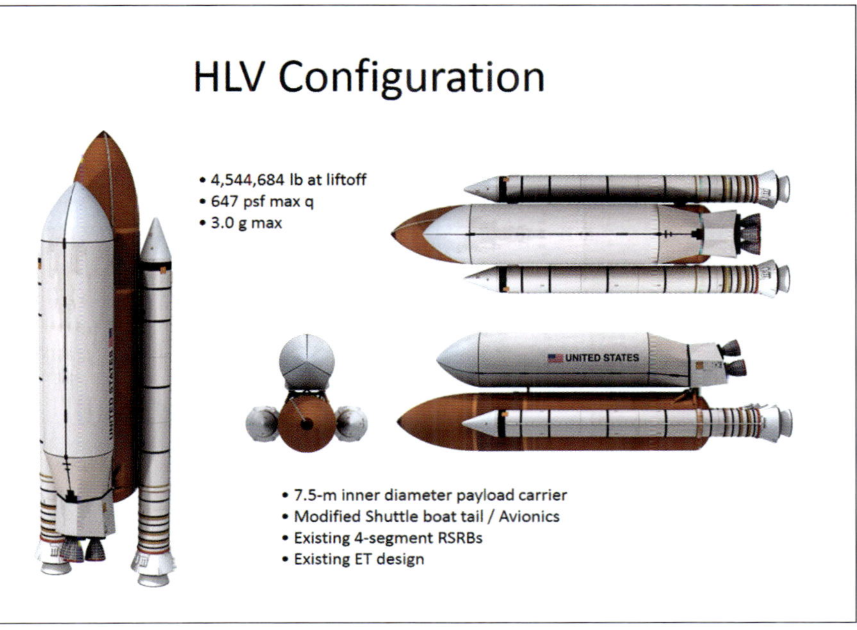

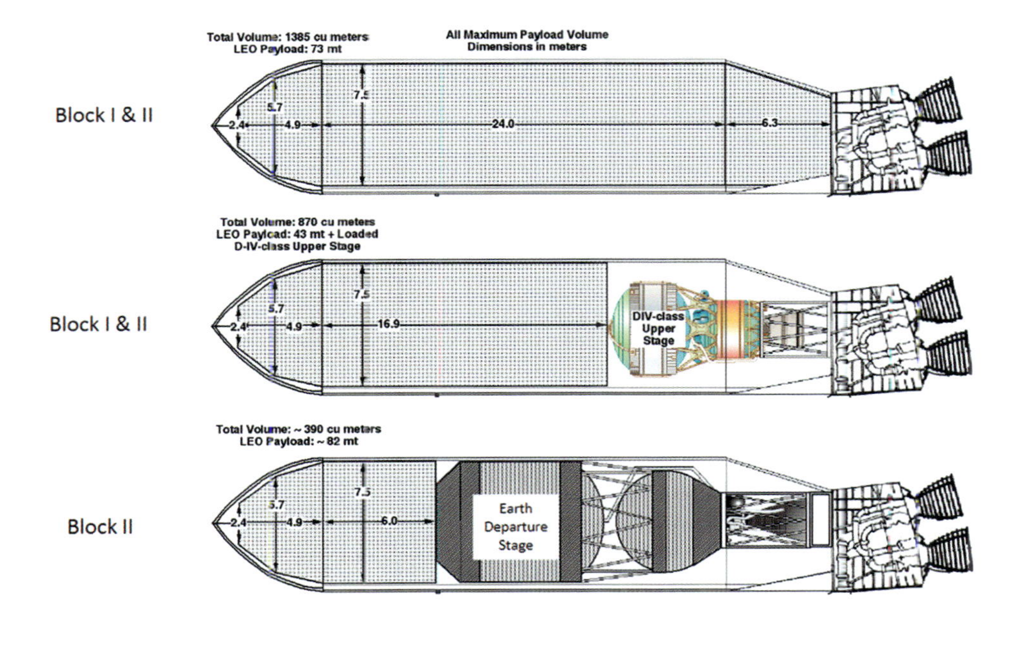

Left: Winged fly-back boosters were considered as long-term development possibilities, but the costs and the reliance on ageing Orbiter hardware made that uneconomical. (NASA)

Below: At the end of the program, detailed analysis showed that at the beginning of flight operations, there had been a 1 in 9 risk of "Loss of Complete Vehicle" (LOCV), reduced to 1 in 90 flights through various improvements and safety measures. Data such as this persuaded NASA to retire the vehicle. (NASA)

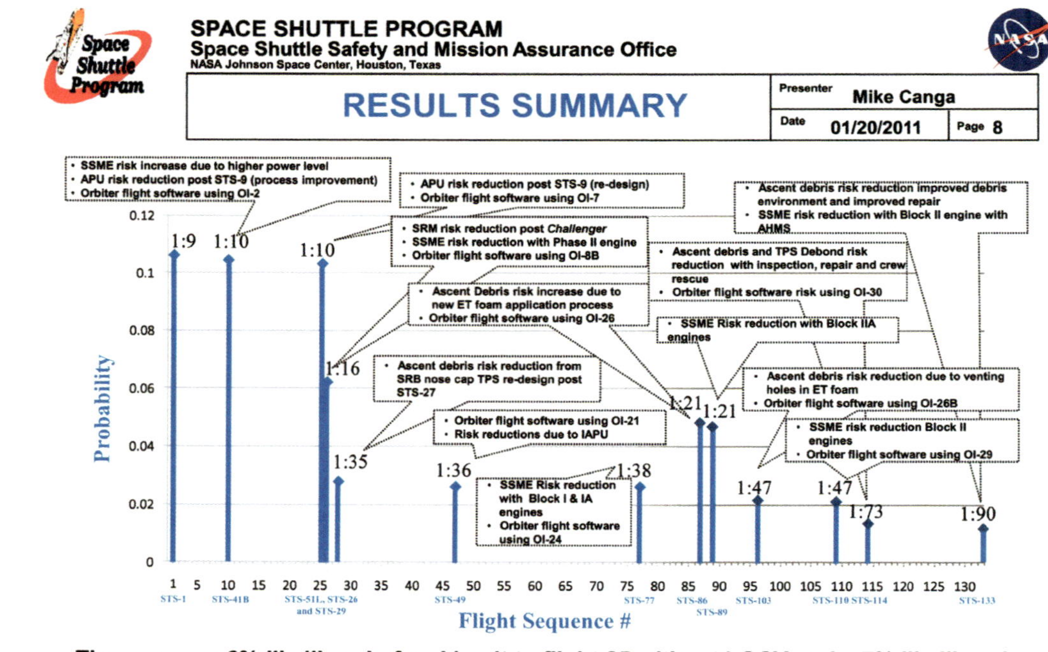

- **There was an 6% likelihood of making it to flight 25 without LOCV and a 7% likelihood of making it from flight 26 to flight 113 without LOCV using the values on this chart**
  - **We were lucky, there were a number of close calls (e.g. STS-9 APU fire, STS-27 Ascent Debris, STS-95 drag chute door)**

Conclusion

The Shuttle has an astonishing record, unlikely to be broken in the foreseeable future. Some 355 people from many countries flew in an Orbiter, but after its retirement in 2011, many lessons were learned about failures in the way the Shuttle had been assessed over time. Only after the program ended could the real evaluation produce definitive results.

While the Orbiters were repeatedly flying missions to space, refurbishment and general maintenance never came close to the unrealistic claim that they could fly weekly or even fortnightly missions. At 39 launches (August 1984–February 2011), *Discovery* had the most flights, followed by *Atlantis* with 33 (October 1985–July 2011), *Columbia* with 28 (April 1981–February 2003), *Endeavour* with 25 (May 1992–May 2011) and *Challenger* with 10 (April 1983–January 1986).

From these figures, it can be seen that the average time between flights with the same Orbiter was around nine months, only *Challenger* achieving the extraordinary figure of 3.3 months on average. But that was due largely to the extreme pressures placed upon the traffic model by the burgeoning flight manifest between 1983 and 1986, and that only two Orbiters were available during that crucial early build-up phase in flight rates.

Being unique in the role for which it was designed, albeit with relatively conventional engineering practices, few had questioned extravagant and outlandish claims about Shuttle flight frequency and its ease of maintenance. This proved much more difficult than expected, and although routine servicing settled into a mature schedule, each mission was unique and brought its own problems. Moreover, no two Orbiters were the same and the weight-reduction effort was considerable with each successive vehicle. Because of that, individual Orbiters were allocated to particular payloads to take advantage of specific capabilities.

Reusability had been a key factor in getting approval for the Shuttle, but that proved troublesome and, in some cases, more expensive. The highly tuned engines required a lot of maintenance; the thermal tiles proved so troublesome during the first phase of Shuttle operations that it pushed engineers to replace them with an alternative protection, as explained in Chapter 2. Fuel cells were efficient, and most of the hydraulic equipment was no more problematic than on a military combat aircraft, but the overall processing was made difficult by the variable nature of each Orbiter and by the lack of spares and replacement parts. Many companies supplying initial equipment and materials had simply gone out of business.

Of those items considered "reusable" the SRBs were the most problematic. The difficulties with recovering them was only a minor problem compared to the damage incurred by salt water, which is never a good environment. It transpired that refurbishment for each SRB cost three times the price of a new booster, and that soured any future concept dependent on splashdown and recovery. This is one reason why SpaceX recovers its Falcon boosters on land, giving up a small percentage of payload to preserve some propellant for decelerating to a soft touchdown.

Was it a good idea to build the Shuttle? Yes, but not for the economic and cost-cutting reasons used to justify it at the outset. What it did do was provide the means of building the ISS, maintain an investment in human space flight, enable the ESA and Japan to gain experience building modules for the ISS, and support Canada's outstanding robotics industry. It also encouraged a new generation of young people in STEM subjects and brought together international partnerships that carry forward into the Artemis program.

Perhaps more than all of that, it enabled 355 people from many countries to experience the "overview effect" of seeing the Earth in all its troubled state as a speck of dust in the vast universe. And with each return came a unanimous call for better respect and coexistence with a highly fragile ecosystem upon which all of life depends – including the future of the human species. For those reasons alone, it was very well worthwhile.

Discovery **launches a payload to the ISS on mission STS-120. (NASA)**

Appendix

Space Shuttle Flights by Orbiter

The following missions are in the chronological sequence flown by order of Orbiter vehicle introduced to the fleet.

ENTERPRISE (OV-101)

Enterprise was used for Air-Launch Flight Tests in 1977 from the top of the converted Boeing 747. It is on display at Intrepid Sea, Air & Space Museum, New York City, New York.

COLUMBIA (OV-102)

Mission	Launch Date	Mission Information
STS-1	Apr 12 1981	1st Shuttle Orbital Flight Test (OFT-1)
STS-2	Nov 12 1981	2nd Shuttle Mission/Office of Space and Terrestrial Applications-1 (OSTA-1)
STS-3	Mar 22 1982	3rd Shuttle Mission/Office of Space Science-1 (OSS-1)
STS-4	Jun 27 1982	Department of Defense, materials experiments
STS-5	Nov 11 1982	First commercial comsats (ANIK C-3 and SBS-C) by Shuttle
STS-9	Nov 28 1983	Orbital Laboratory and Observations Platform/First Spacelab mission
STS-61C	Jan 12 1986	SATCOM KU-1 comsat
STS-28	Aug 8 1989	Two Department of Defense comsats (DSCS-III F-2 and F-3)
STS-32	Jan 9 1990	SYNCOM IV-F5 comsat; LDEF retrieval (see *Challenger* STS-41C)
STS-35	Dec 2 1990	Spacelab ASTRO-1 astronomy mission
STS-40	Jun 5 1991	Spacelab Life Sciences-1
STS-50	Jun 25 1992	Spacelab microgravity research USML-1
STS-52	Oct 22 1992	USMP-1, LAGEOS II
STS-55	Apr 26 1993	Spacelab D-2 German mission
STS-58	Oct 18 1993	Spacelab SLS-2 life sciences
STS-62	Mar 4 1994	USMP-2, OAST-2 science mission
STS-65	Jul 8 1994	Spacelab IML-2 international microgravity mission
STS-73	Oct 20 1995	Spacelab microgravity research USML-2
STS-75	Feb 22 1996	Tethered satellite (TSS-1R), US Microgravity Payload-3 science mission
STS-78	Jun 20 1996	Spacelab life and microgravity (LMS) science
STS-80	Nov 19 1996	Free-flying science satellites (ORFEUS-SPAS II and WSF-3)
STS-83	Apr 8 1997	Spacelab MSL-1 material science mission
STS-94	Jul 1 1997	Spacelab MSL-1 material science re-flight mission

Mission	Launch Date	Mission Information
STS-87	Nov 19 1997	US Materials Processing-4 mission; Spartan 201-04 free-flying satellite
STS-90	Apr 17 1998	Final Spacelab mission, neurological sciences mission
STS-93	Jul 23 1999	Chandra X-Ray Observatory
STS-109	Mar 1 2002	Hubble Space Telescope servicing mission
STS-107 *	Jan 16 2003	Microgravity research mission/SPACEHAB

* *Columbia* and crew were lost on February 1, 2003, during re-entry over East Texas, approximately 16 minutes before landing.

CHALLENGER (OV-99)

Mission	Launch Date	Mission Information
STS-6	Apr 4 1983	Tracking and data relay satellite-1 (TDRS-1)/1st Shuttle space walk
STS-7	Jun 18 1983	Indonesian and Canadian comsats/1st US woman in space
STS-8	Aug 30 1983	Multipurpose satellite launch for India/1st night launch and landing
STS-41B	Feb 3 1984	US and Indonesian comsats, Manned Manoeuvering Unit, 1st KSC landing
STS-41C	Apr 6 1984	Long Duration Exposure Facility deployed (LDEF), 1st on-orbit spacecraft repair
STS-41G	Oct 5 1984	US Earth Radiation Budget Satellite (ERBS)
STS-51B	Apr 29 1985	Spacelab-3 microgravity and life sciences mission
STS-51F	Jul 29 1985	Spacelab-2 solar physics mission
STS-61A	Oct 30 1985	Spacelab D-1 (1st German dedicated Spacelab)
STS-51L**	Jan 28 1986	TDRS-2 satellite, SPARTAN-203 free-flyer

***Challenger* and crew were lost on January 28, 1986, approximately 73 seconds after lift-off, following a catastrophic explosion in the right solid rocket booster.

DISCOVERY (OV-103)

Mission	Launch Date	Mission Information
STS-41D	Aug 30 1984	SBS-D; TELSTAR 3C, SYNCOM IV-2 comsats; Solar Wing experiment
STS-51A	Nov 8 1984	Canadian comsat Anik D2, Leasat-1 US Navy comsat
STS-51C	Jan 24 1985	Orion 1 signals intelligence satellite for Defense Department
STS-51D	Apr 12 1985	Canadian comsat Anik C3, Leasat-3 US Navy comsat
STS-51G	Jun 17 1985	MORELOS-A, ARABSAT-A and TELSTAR-3D comsats
STS-51I	Aug 27 1985	US ASC-1 satellite, Australia's AUSSAT-1, Leasat-2 US Navy comsat
STS-26	Sept 29 1988	TDRS-C launch
STS-29	Mar 13 1989	TDRS-D launch
STS-33	Nov 22 1989	Orion 2 signals intelligence satellite for the Department of Defense
STS-31	Apr 24 1990	Hubble Space Telescope deployed
STS-41	Oct 6 1990	Ulysses spacecraft to Jupiter, several science experiments

Mission	Launch Date	Mission Information
STS-39	Apr 28 1991	Department of Defense experimental research mission
STS-48	Sept 12 1991	Upper Atmosphere Research Satellite
STS-42	Jan 22 1992	Spacelab IML-01 microgravity mission
STS-53	Dec 2 1992	SDS B3 comsat on last dedicated Department of Defense mission
STS-56	Apr 8 1993	ATLAS-2 earth atmosphere science mission, free-flying SPARTAN-201
STS-51	Sept 12 1993	Comsat technology satellite, astronomy science (ORFEUS-SPAS)
STS-60	Feb 3 1994	Free-flying physics experiment (WSF-1) SPACEHAB-2
STS-64	Sept 9 1994	Physics experiment LITE, and free-flying SPARTAN-201
STS-63	Feb 3 1995	SPACEHAB-3 materials and life sciences mission
STS-70	Jul 13 1995	TDRS-G
STS-82	Feb 11 1997	2nd Hubble Space Telescope servicing
STS-85	Aug 7 1997	Earth science mission (CRISTA-SPAS-02)
STS-91	Jun 2 1998	9th and final Shuttle-*Mir* docking
STS-95	Oct 29 1998	John Glenn's flight; SPACEHAB
STS-96	May 27 1999	2nd International Space Station flight
STS-103	Dec 19 1999	3rd Hubble Space Telescope servicing mission
STS-92	Oct 11 2000	International Space Station assembly flight 3.3A
STS-102	Mar 8 2001	International Space Station assembly flight 5A.1
STS-105	Aug 10 2001	International Space Station assembly flight 7A.1
STS-114	Jul 26 2005	International Space Station assembly flight LF1
STS-121	Jul 4 2006	International Space Station assembly flight ULF1.1
STS-116	Dec 9 2006	International Space Station assembly flight 12A.1
STS-120	Oct 23 2007	International Space Station flight 10A
STS-124	May 31 2008	International Space Station flight 1J
STS-119	Mar 15 2009	International Space Station flight 15A
STS-128	Aug 28 2009	International Space Station flight 17A
STS-133	Feb 24 2011	International Space Station assembly flight ULF5
Discovery is displayed at the Steven F. Udvar-Hazy Center in Chantilly, Virginia.		

ATLANTIS (OV-104)

Mission	Launch Date	Mission Information
STS-51J	Oct 3 1985	Department of Defense communications satellite mission
STS-61B	Nov 26 1985	Mexican (MORELOS-B), Australian (AUSSAT-2), US SATCOM KU-2
STS-27	Dec 2 1988	Lacrosse radar imaging satellite for Department of Defense
STS-30	May 4 1989	Magellan spacecraft to Venus
STS-34	Oct 18 1989	Galileo; Shuttle Solar Backscatter Ultraviolet Experiment
STS-36	Feb 28 1990	Mysty reconnaissance satellite for Department of Defense
STS-38	Nov 15 1990	SDS-B2 comsat for Department of Defense

Mission	Launch Date	Mission Information
STS-37	Apr 5 1991	Gamma Ray Observatory
STS-43	Aug 2 1991	TDRS-E, physics experiments
STS-44	Nov 24 1991	Early warning satellite (DSP-F-16) for Department of Defense
STS-45	Mar 24 1992	Atmospheric sciences mission (ATLAS-1)
STS-46	Jul 31 1992	Tethered satellite (TSS-1), EURECA free-flyer deployed
STS-66	Nov 3 1994	Atmospheric sciences mission (ATLAS-3), CRISTA-SPAS
STS-71	Jun 27 1995	1st Shuttle-*Mir* docking
STS-74	Nov 12 1995	2nd Shuttle-*Mir* docking
STS-76	Mar 22 1996	3rd Shuttle-*Mir* docking; SPACEHAB
STS-79	Sept 16 1996	4th Shuttle-*Mir* docking
STS-81	Jan 12 1997	5th Shuttle-*Mir* docking
STS-84	May 15 1997	6th Shuttle-*Mir* docking
STS-86	Sept 25 1997	7th Shuttle-*Mir* docking
STS-101	May 19 2000	3rd International Space Station flight
STS-106	Sept 8 2000	International Space Station Flight 2A.2b
STS-98	Feb 7 2001	International Space Station assembly flight 5A
STS-104	Jul 12 2001	International Space Station assembly flight 7A
STS-110	Apr 8 2002	International Space Station 8A
STS-112	Oct 7 2002	International Space Station 9A
STS-115	Sept 9 2006	International Space Station assembly flight 12A
STS-117	Jun 8 2007	International Space Station assembly flight 13A
STS-122	Feb 7 2008	International Space Station/ESA Columbus laboratory
STS-125	May 11 2009	Hubble Space Telescope final servicing
STS-129	Nov 16 2009	International Space Station ULF3
STS-135	Jul 8 2011	International Space Station Multi-Purpose Logistics Module
Atlantis is now displayed at the Visitor Complex at the Kennedy Space Center, Florida.		

ENDEAVOUR (OV-105)

Mission	Launch Date	Mission Information
STS-49	May 7 1992	Intelsat VI repair
STS-47	Sept 12 1992	Spacelab-J dedicated Japanese experiments
STS-54	Jan 13 1993	TDRS-F, X-ray spectrometer
STS-57	Jun 21 1993	SPACEHAB-1; EURECA retrieval
STS-61	Dec 2 1993	1st Hubble Space Telescope servicing mission
STS-59	Apr 9 1994	Space Radar Laboratory (SRL-1)
STS-68	Sept 30 1994	Space Radar Laboratory (SRL-2)
STS-67	Mar 2 1995	Ultraviolet Observatory science mission (ASTRO-2)

Mission	Launch Date	Mission Information
STS-69	Sept 7 1995	Free-flyer SPARTAN 201-03, physics experiment (WSF-2)
STS-72	Jan 11 1996	Science and technology mission (SFU), OAST-Flyer
STS-77	May 19 1996	SPACEHAB; SPARTAN (IAE)
STS-89	Jan 22 1998	8th Shuttle-*Mir* docking
STS-88	Dec 4 1998	1st International Space Station flight
STS-99	Feb 11 2000	Shuttle Radar Topography Mission
STS-97	Nov 30 2000	International Space Station assembly flight 4A
STS-100	Apr 19 2001	International Space Station assembly flight 6A
STS-108	Dec 5 2001	International Space Station assembly flight UF-1
STS-111	Jun 5 2002	International Space Station UF2
STS-113	Nov 23 2002	International Space Station 11A
STS-118	Aug 8 2007	SPACEHAB; ISS CMG replacement
STS-123	Mar 11 2008	Japanese Kibo module & Canadian manipulator (Dextre) to ISS
STS-126	Nov 14 2008	International Space Station upgrade
STS-127	Jul 15 2009	International Space Station Japanese Kibo assembly
STS-130	Feb 8 2010	International Space Station assembly
STS-134	May 16 2011	International Space Station assembly flight ULF6
Endeavour is on display at the California Science Center, Los Angeles, California.		

Other books you might like:

Historic Commercial Aircraft
Series, Vol. 12

Historic Military Aircraft
Series, Vol. 17

For our full range of titles please visit:
shop.keypublishing.com/books

VIP Book Club

Sign up today and receive
TWO FREE E-BOOKS

Be the first to find out about our forthcoming
book releases and receive exclusive offers.

Register now at keypublishing.com/vip-book-club